Diaspora and Disaster

Japanese Outside Japan and the Triple Catastrophy of March 2011

d|u|p

Kultur- und Sozialwissenschaftliche Japanforschung

Band 1

Diaspora and Disaster

Japanese Outside Japan and the Triple Catastrophy of March 2011

herausgegeben von

Andreas Niehaus und Christian Tagsold

d|u|p

Bibliografische Information der Deutschen Nationalbibliothek
Die Deutsche Nationalbibliothek verzeichnet diese Publikation in der
Deutschen Nationalbibliografie; detaillierte bibliografische Daten sind
im Internet über http://dnb.dnb.de abrufbar.

© düsseldorf university press, Düsseldorf 2016
http://www.dupress.de
Redaktion und Lektorat: Jutta Teuwsen
Umschlaggestaltung: Hannah Reller
Satz: Jutta Teuwsen, LaTeX

Gesetzt aus der Minion Pro und der URW Classico
ISBN 978-3-95758-005-4

Contents

Diaspora and Disaster
3/11 as a Starting Point of Connecting both Fields of Research

This volume brings together two topics and frameworks of research which up to now have rarely been connected. First a distinct "disasterology" has dealt with the preconditions and consequences of disaster[1], and especially social anthropology has contributed strongly to a deeper understanding of disasters as *social* catastrophes. Second we find a fashionable and not as well defined field of research in diaspora studies. Diaspora studies have been taken up by social and cultural anthropology, literature studies, sociology and many more in the humanities. Only lately have both fields of disaster research and diaspora studies been treated together by a few papers as in the case of Haitians in the USA trying to come to terms with the Haiti earthquake in 2010 (Esnard and Sapard 2011). These papers have proven the potential of grouping disaster and diaspora together, but have only partially lived up to the possibilities that lie within this approach. With our volume on the triple catastrophe in Northern Japan in March 2011 and the reaction of Japanese abroad we try to move one step further.

The research fields of disaster and diaspora both have certain blind spots which are somehow complimentary to each other. First, disaster research tends to think along national borders. As a consequence disasters are analyzed as national catastrophes.[2] It is true that nation-states form the most efficient institutions to react to disasters unless the state has been very weak already before. They are responsible for preparation and mitigation of disasters or the failure in these two dimensions. Since disasters are not natural catastrophes as media often claims but are conditioned by social circumstances (Bakoff, Frerks, Hillhorst 2004: 1) it is evident that the state should be the starting point for analysis.

---

[1] Especially Oliver-Smith (1986, 1999, 2004) has been a leading researcher of disaster in social anthropology while others like Alexander (1997) have also made important contributions to the research of disasters as social-natural catastrophes.

[2] For example Alexander (1997: 25) claims: "In a sense, a disaster is symptomatic of the condition of a society's total adaptional strategy within its social, exonomic, modified, and build environments." He thus supposes that disasters are only symptoms for the condition of the very society hit by them.

In the case of 3/11, as the triple catastrophe is often referred to as the initial earthquake occurred on March 11th 2011 and also because the term resembles 9/11, Japan was basically well prepared to meet natural catastrophes like earthquakes, taifuns and tsunamis. Experts estimated the probability of a major tsunami caused by an earthquake on the coast of Sanriku, the name for the north-eastern region of the main island Honshū, to be around 90 % (Tagsold 2013a: 609). Accordingly inhabitants knew how to react and practiced how to most effectively evacuate. However, the actual tsunami's dimensions exceeded all scientific predictions. Nevertheless the Japanese state proved its resilience during March 2011 as the impact of the natural disaster in terms of casualties as well as destruction would have much more severe in most other countries. However the third catastrophe, the meltdown at the nuclear power plant Fukushima Daiichi came unexpectedly and led to chaos.

Up to this, disasterology has treated 3/11 mostly as a purely Japanese matter and thus has not enlightened the blind spot by extending its research beyond national borders of Japan.[3] Books and papers have dealt with the various dimensions of the disaster like living in shelters, food safety, reconstruction of cities, ports and infrastructure in general and of course Fukushima. But the disaster has a strong transnational dimension, which has more or less been neglected. First of all Northern Japan housed a significant number of migrants in March 2011 (Tagsold 2013b: 182f.). Even though migration to Japan always has been seriously curbed by the government, the country does not seclude itself from the trends of globalization. Chinese nationals worked on the farms in rural Northern Japan, as Japanese tend to leave for the big cities since wages on the countryside are low and work is hard and dirty. Nevertheless, these jobs seemed to offer opportunities for migrants. On the coast many Filipinas had married fishermen who otherwise faced difficulties finding partners for marriage in Japan. These migrants and others have widely been neglected in studies concerning the Triple Disaster, as somehow the assumption as well as public image is that the disaster had hit a Japanese island that is populated solely by Japanese.

---

[3] See for examples the papers in Kingston (2012).

Yet another transnational dimension of the disaster is constituted by the Fukushima nuclear power plant. One of the common implicit assumptions in public media was that the Fukushima power plant is a purely Japanese achievement. Hence everything regarding these nuclear reactors was explained by referring to some Japanese national character. The Japanese have full faith in technical solutions to overcome even the most severe disasters. But the walls, which were built to stop tsunami-waves, failed – and so too did the power plant in Fukushima. Nature had overcome human – that is Japanese – hybris, so it was argued. As far as the power plants are concerned however, research has clearly outlined their transnational background (see for example Yoshioka 1999). The US advertised nuclear power as safe and as the energy of the future in the 1950s and 60s in Japan in order to strengthen one of their last reliable allies in East Asia (Kuznick 2011, Tanaka 2011). The multinational company General Electrics built the first nuclear reactor in Fukushima and cooperated with Toshiba in building two additional reactors, only the last three reactors were built solely by Japanese companies. But the transnational networks are not only historical – they also reached out globally in March 2011, a point less explored by experts on Japanese nuclear power politics. For example the reaction of the US marines moving their ships away from Fukushima or closer back was one of the often quoted events in media to determine how dangerous the incident might be. Also the International Atomic Energy Agency in Vienna played an important role in defining the disaster in Fukushima. Even though studies of the disaster partly grasped the transnational character of 3/11 for Fukushima they nevertheless fell short in consequently asking the same questions about the disaster.

Diaspora studies have far less problems with transnational dimensions. The framework of analysis is based on the very question of overcoming national boarders. After all, as Nico Besnier in his introductory paper stresses, diaspora studies have their own blind spot. There is a tendency to treat the presence of people with the same foreign background in a city or a country as a distinct "community". However the notion of community evokes a sense of shared values, practices and methods of identifying oneself with the alleged country of origin, which often runs counter to the real degree of cohesiveness among these people. A working infrastructure for one group of citizens like a Chinatown or Little Italy might seem to prove a shared identity among these citizens in a foreign and yet homey country. Still, the mere existence of food shops, bookstores, or even a consulate general doesn't say much about identities of the people termed "members" of the community. This is certainly true for Japanese diasporas worldwide which are often supported by a Japanese infrastructure and are advertised as coherent groups. However, most Japanese have much more complex strategies for identifying themselves than advertising themself as a personification of a perfect Japanese; some do at times for various reasons, others refrain from doing so.

In contrast to this understanding of diaspora we look for performative moments in which citizens identify themselves as diasporic. A cultural center or a festival which normally would be seen as an expression of identity of an diasporic "community" thus turns into a moment in which diaspora and even community is created – momentarily and dynamic, not as a static assumption about "belonging". 3/11 certainly was such a moment in which many people worldwide were emotionally hit by the dramatic situation in Northeast Japan and felt the need to show their solidarity. Various motives drove these people. A basic human compassion, a long standing history of personal protest against nuclear power, personal experience of similar disasters or a sympathy for Japan and the Japanese might have been reasons for donating, starting relief projects and other ways of helping. Another strong motive for many people to engage themselves into relief projects for North Japan was their Japanese descent and their feeling that they had to identify with the "country of their origin". However, in a sense, terminology is misleading here and in itself suggests "natural bonds" between Japanese abroad and Japan. The question of origins only became virulent for some people because of 3/11 and their feeling that they had to react somehow. Others did not feel any obligation and thus did not stress origins.

In our volume we look at the effects of 3/11 on diasporas in various places. First Nico Besnier's article concentrates on frameworks and concepts for dealing with diaspora. His theoretical introduction forms the theoretical setting for the following papers as it challenges the idea of diaporas as tightly knit *community*.

Based on this critique we scrutinize reactions to 3/11 and ask about the motives of the various actors in the unfolding process. Next our volume analyses reactions to two different types of Japanese diasporas as outlined by Harumi Befu (2001: 12). Japanese diasporas in Düsseldorf, Belgium and London belong to the rotating type according to Befu. Most Japanese are not permanently in Europe, they have been dispatched by the companies for a period of three to five years or are family members of dispatched managers. This group's ties to Japan are mostly very strong and absolutely unquestioned, and the 3/11 disaster was a test for them regarding the relation of the Japanese to their surroundings. Ruth Martin shows how this led to overwhelmingly positive reactions and feelings in London. The Japanese living there were successful in organizing meaningful public events which helped to console the grievance and strengthen ties to non-Japanese. In contrast the public events of mourning staged in Düsseldorf – the location of the third largest Japanese diaspora in Europe after London and Paris – were somewhat unsettling and unfolded their meaning only in the coverage of the local media but less so for those taking part as Christian Tagsold shows in his paper. Finally Andreas Niehaus and Tine Walravens present the findings of their research into the reaction of long-term and short-term Japanese nationals living in Belgium following the Triple Disaster.

Jutta Teuwsen and Peter Bernardi present findings on two permanent diasporas. Teuwsen asks about the meaning of 3/11 for Japanese in Hawaii. Most Japanese came to the islands about a century ago. The current diaspora is made up mostly of third-, fourth- and fifth-generation Japanese. For a long time these Japanese-Americans have avoided identifying themselves by the history of migration of their (great) grandparents, due to Japan's aggression in World War II. However, 3/11 sparked a new trend of identification with Japan. As Bernardi shows, Japanese in Sao Paolo have never felt the need to disguise their migrational history. After 3/11 various actors have tried, however, to lead the broad movement of solidarity and channel donations through their organization to gain reputation and recognition in Brazil as well as in Japan.

Literature

ALEXANDER, David. 1997. "The Study of Natural Disasters, 1977–97: Some Reflections on a Changing Field of Knowledge." *Disasters* 21 (4): pp. 284–304.

BANKOFF, Greg; FRERKS, George; HILHORST, Dorthea (eds.). 2004. *Mapping Vulnerability: Disasters, Development, and People.* London, Sterling, VA: Earthscan Publications.

BEFU, Harumi. 2001. "The Global Context of Japan Outside Japan." In: Befu, Harumi; Guichard-Anguis, Sylvie (eds.): *Globalizing Japan: Ethnography of the Japanese Presence in Asia, Europe and America.* London: Routledge, pp. 3–22.

ESNARD, Ann-Margaret; SAPAT, Alka. 2011. "Disasters, Diasporas and Host Communities: Insights in the Aftermath of the Haiti Earthquake." In: *Journal of Disaster Research* 6 (3): pp. 331–342.

KINGSTON, Jeff (ed.). 2012. *Natural Disaster and Nuclear Crisis in Japan: Response and Recovery After Japan's 3/11.* London, New York: Routledge.

KUZNICK, Peter. 2011. "Japan's Nuclear History in Perspective: Eisenhower and Atoms for War and Peace." In: *The Asia-Pacific Journal* 9 (18), http://www.japanfocus.org/-yuki-tanaka/3521. (accessed November 25, 2015)

OLIVER-SMITH, Anthony (ed.). 1986. *Natural Disasters and Cultural Responses.* Williamsburg, Va.: Dept. of Anthropology, College of William and Mary.

OLIVER-SMITH, Anthony; HOFFMAN, Susannah M. (eds.). 1999. *The Angry Earth: Disaster in Anthropological Perspective.* New York: Routledge.

OLIVER-SMITH, Anthony. 2004. "Theorizing Vulnerability in a Globalized World: A Political Ecological Perspective." In: Bankoff, Greg; Frerks, George; Hilhorst, Thea (eds.): *Mapping Vulnerability: Disasters, Development, and People.* London, Sterling, VA: Earthscan Publications. pp. 10–24.

TAGSOLD, Christian. 2013a. "Die dreifache Katastrophe vom März 2011 und die alternde Gesellschaft in Tōhoku." In: *Asiatische Studien* 67 (2): pp. 601–624.

TAGSOLD, Christian. 2013b. *Japan: Ein Länderporträt.* Berlin: Ch. Links Verlag.

TANAKA, Yuki. 2011. "'The Peaceful Use of Nuclear Energy': Fukushima and Hiroshima." In: *The Asia-Pacific Journal* 9 (18), http://www.japanfocus.org/-yuki-tanaka/3521. (accessed November 25, 2015)

YOSHIOKA, Hitoshi. 1999. "Forming a Nuclear Regime and Introducing Commercial Reactors." In: Nakayama, Shigeru; Gotō, Kunio; Yoshioka, Hitoshi (eds.): *A Social History of Science and Technology in Contemporary Japan: Road to Self-Reliance, 1952 – 1959.* Melbourne: Trans Pacific Press, pp. 80–124.

Diasporas: Communities of Practice, Economies of Affect

Niko Besnier

It is at times of extreme public trauma that concepts like "community", "identity" and "cohesion" become the object of sustained attention. At the same time, this attention tends to be based on a naturalization of the concepts, the assumption that, on the one hand, these categories do not need to be defined or analyzed and, on the other hand, that they are morally positive, essentially good and important goals for all to pursue. What I would like to do in this brief intervention is to examine some of these assumptions and suggest ways in which we can better theorize our understanding of social action at times of intense trauma in the context of transnational dispersal and diverse connections.

Perhaps a fruitful point of departure is an examination of the oft-cited and seldom questioned concept "identity". Its genealogy in the social sciences encapsulates the complexities and contradictions that are embedded in the very concept itself. We generally associate the birth of a social scientific understanding of identity with developmental psychologist and psychoanalyst Erik Erikson's pivotal work, *Identity and the Life Cycle* (Erikson 1959). In this work, Erikson sought to capture the insight that, while people strive to maintain to themselves and to others a sense of continuity and growth, they go through life facing numerous and diverse challenges. As the challenges they face become more complex and diverse, people's sense of self increases in complexity. While Erikson focused primarily on the development of a psychological identity, he had also been strongly influenced by North American cultural anthropologists of his time; thus he recognized that persons are embedded in social groups and stressed that their sense of identity is as much the product of the people around them as the outcome of intra-personal development. Continuity in the context of change, person-internal as well as person-external, and the fragile play of divergent forces were integral to the concept of identity since its timid emergence in our analytic vocabulary, and militates for an approach to identity as a contingent, interactive and unfinished project. It has also motivated some scholars

Niko Besnier

to argue that the concept is so fraught with problems (e. g. its instability as an analytic and experiential category) to completely undermine its usefulness or to replace it with another, more useful category like subjectivity (Brubaker et al. 2000: 1-47; Ortner 2005: 31-52).

Just as fragile and complex is our notion of "community", which of course goes much further in time than "identity", originating in the social sciences in the visions of Alexis de Tocqueville, Ferdinand Tönnies and Émile Durkheim (if not before). It is coming to us via a century's worth of sociological and anthropological thinking and writing, to late twentieth-century reclaimings of community in some social scientific and philosophical circles, as in the works of Robert Bellah and his colleagues Charles Taylor and Christopher Lasch (Bellah et al. 2007; Taylor 1989; Lasch 1979). Community and identity bear striking resemblances to one another; equally unstable, the two concepts are the subject of a great deal of romanticization of an idealized past in which social cohesion was made unproblematic by feelings of sameness (cf. both Durkheim and Karl Marx), as well as an idealized present and future, in which identity politics and the longing for community reify and fetishize them. Just as community is relentlessly portrayed as an "unequivocal good, an indicator of a high quality of life, a life of human understanding, caring, selflessness, belonging", identity is for many a must and forms the basis of claims to recognition, pushes for legal protection and personal feelings of pride and achievement (Joseph 2002: vii).

The relationship between identity and community, on the one hand, and what they are supposed to oppose, on the other, is fraught. Identity politics, based at once on the construction of sameness and difference, seeks to rectify past and present injustices in the form of oppression, denial of rights, and non-recognition. However, identity politics is also based on reified notions of authenticity and romanticized understandings of community. In fact, many invocations of community in the name of progressive causes and social justice fail to understand the extent to which they themselves are produced by the very source of inequality and lack of justice in society, namely capitalism (Against et al.2006: 3-22). Witness, for example, the extent to which racial, ethnic, sexual, and other forms of social differentiation that form the basis of a subaltern politics of liberation from

14

oppression and marginalization are understood through acts of consumption, through the discrimination between those who belong and those who fail to belong, and through acts of power within the ranks of discriminated groups. Thus, for example, African-American identity politics in the United States (where identity politics was essentially invented) becomes co-opted by the consumption of "Afrocentric" products, from textiles to music to names (Stoller 2002; Boateng 2004: 212-226). And nowhere is the importance of "who belongs and who does not" more strident than in criticisms of the current president of the United States for not being "black enough" by some and "not American" by others.

Now (as if all this were not complicated enough already), transnational movement creates additional complexities, particularly when this movement takes the shape of what we have come to call "diasporic dispersal". So it is to diasporas that I now turn. As is well known, the term itself is from Old Testament Greek. It was originally borrowed into English in the late-nineteenth century. (The *Oxford English Dictionary* lists a first occurrence in 1876, and it appears in the 1881 edition of the Encyclopædia Britannica under the entry for "Israel".) For the next century, the term referred almost exclusively to the successive dispersals of Jews in the ancient world. It was only in the 1990s that the term entered social scientific language, where it became associated with sociological and anthropological approaches to people's mobility that defied a simplistic understanding of migrations as movement from point A to point B, followed by the relative integration of migrants into the social conditions of point B and their gradual disengagement from the social conditions of point A. At the time, a drive was afoot in understanding the movement of people not as a matter of "migration", which conforms to this simplistic model, but as "mobility" (Hannam et al. 2006: 1-22; Urry 2007). This considerably more dynamic concept allows for movement between not just two but multiple points, not only unidirectional movement but also backtracking and sidestepping, as well as a play of multiple allegiances and senses of belonging, all of which, if we believe globalization theorists, are new to the late twentieth century.

Diasporic situations involve all of these features, with a particular emphasis on a number of characteristics (Brubaker 2005: 1-19). The first and perhaps least controversial defining feature of a diaspora is dispersal to multiple destinations. In contrast to migrant populations following a predictable path determined by clear historical contingencies (e. g., Algerians moving to France after World War II), diasporas consist of people moving to different geographical points. The diaspora on which my own work focuses is a good example: since the 1960s, Tongans have moved in relatively large numbers from their island kingdom in the South Pacific to New Zealand, Australia, Hawaii, and the Continental United States, as well as, in smaller numbers, to a vast list of destinations, including Japan, to the extent that very few countries of the world fail to count at least a small number of Tongans in residence. The effect is one of an "exploded" population, one that moves in the multiple points of the compass from their point of origin (Besnier 2011; Small 2011; Lee 2003). What plays a determinative role in the relative multiplicity of destinations are state borders and the legal and punitive regimes associated with them.

The second feature that characterizes diaspora is what Rogers Brubaker calls "boundary maintenance", namely the work that people do to distinguish themselves from those who are outside the group and by implication the work that they do to emphasize commonality and homogeneity within the group. Here, Brubaker bases himself on Norwegian anthropologist Fredrik Barth's classic and yet still profoundly relevant thesis that ethnicity is not the product of the internal properties of a group (e. g., symbols, physical appearance, practices, objects) but the product of action designed to distinguish the group from those outside the group with whom they come into contact (Barth 1969: 9-38). For Brubaker, diasporic citizens "work" to distinguish themselves from those around them and do so through various means: endogamy ("marry only someone from your group"), for example, as well as other forms of self-segregation.

In my own view, this is the weakest factor because it fails to distinguish the ideology of being different from the social practice of self-differentiation. Take endogamy in the context of the most canonical diaspora of all, namely the Jewish diaspora over the centuries. As is well known, for many centuries Jews in Europe held on to a strong ideology of endogamy and other forms of separation from mainstream society, and of course mainstream society returned the favor by excluding Jews from institutions, spaces and structures of power, or by attempting to eliminate them. However, in his explosive book, *The Invention of the Jewish People*, originally published in Hebrew in 2008, Israeli historian Shlomo Sand has argued not only that European and Middle Eastern Jews were in constant and often cordial contact with their non-Jewish neighbors (something that was certainly the case in the multicultural Ottoman Empire that ruled over the Middle East for centuries) but also that most modern Jews descend from converts and inter-marriages (Sand 2009 and Wayland 2004: 405-426). It is only in the nineteenth century that Jewish intellectuals, inspired by Romantic constructions of the folk character of German nationalism, began constructing a history of the Jews as a wandering and separate people who would eventually return to the Promised Land. (As one can well imagine, the book created an uproar because it essentially destroys the basis of Zionist claims to land in Israel and Palestine and everything that stands on it.) For the purpose of our understanding of diasporas, it modulates the importance of a historically continuous sense of otherness as a necessary feature of diasporas.

The third and perhaps most interesting feature (as well as more directly relevant to the situation that this special issue focuses on) is a sustained orientation to the homeland, be it real or imagined. What "orientation" actually means remains vague in many representations. In the example that I have already provided, that of the Tongan diaspora, as well as many other cases like it, "orientation" is commonly measured or evaluated in terms of material support: for example, the classic "remittances" or sometimes substantial sums of money that mobile citizens send to their relatives, their churches or mosques, their villages, political parties or armies and militias (witness the support of the Tamil diaspora for the Tamil Tigers during the civil war in Sri Lanka) (Wayland 2004: 405-426).

In fact, the motivation for moving in many cases is predominantly the need and moral responsibility to support non-mobile co-citizens. These feelings of responsibility are often embedded in long-term structures of reciprocity and indebtedness, from the religious responsibility that Buddhism places on Thais to support their elders to the classic Maussian counter-gift that migrants owe to their families (or, in darker contexts, to human traffickers) who have financed their travels. This situation, of course, is much more relevant to diasporas that emanate from homelands where money is in short supply, employment is lacking, and, even if one finds employment, the income one generates even in high-level local employment is considerably lower than the income one generates in low-level employment elsewhere. Thus, Filipino medical doctors find it preferable to work as nurses in Canada (where their medical doctors' qualifications are not recognized) than to work as doctors in the Philippines, despite the downward economic mobility that this decision represents (McElhinny et al. 2009: 93-110).

But there are other ways in which "orientation to a homeland" can manifest itself, and this is what is relevant here. If we replace "orientation" with "allegiance", namely the loyalty or commitment of a person to a group or a cause, then we can begin to encompass a broader range of diasporic situations, including the one focused in the present work. But, at the same time, allegiance is a slippery notion. What is the group or cause that I feel allegiance to? When I sent some money to the Red Crescent to help people who have lost loved ones and property in the Pakistani floods of 2010, am I expressing allegiance to a group or a cause, or simply expressing a humanistic sense of empathy for fellow human beings, no matter where they live, what religions they practice, and what conditions they live in? But is my act different than the money I sent to the Japanese Red Cross on March 2011, because of my much greater personal involvement with Japan than with Pakistan? Furthermore, allegiance can be complex, in that I can feel deep sorrow and empathy for people whose families and homes were obliterated by the tsunami, but how do I feel about the cronyism between state power and industrial power, colluding to keep Japanese people in the dark as to the real danger of radiation? Clearly, allegiance must be to something, and that something may vary greatly from one person to the other, from one group to the other, and from one moment to the next. That "something" certainly does not mean "country", a concept that conflates nation, state, the corporate world, and the kind of identity that the nihonjinron ("scholarship of Japanese uniqueness") industry would like us to believe operates in Japan. What these considerations imply is that feelings and action are closely bound together, and that people define themselves, define who they are in the context of a transnational "community", and by implication define what a diaspora is, in terms of "affect", a category that seeks to capture an essentially psychological phenomenon (otherwise known as "emotion" or "feeling") while emphasizing the extent to which this phenomenon is the motivation for action in response to other actions. The close relationship between affect and action helps us rethinking diasporic communities not so much as classic instances of Durkheimian or Tönniesian "communities", or even as instances of "imagined communities" that Benedict Anderson sees at the root of nationalism, but in terms of what has been termed "communities of practice" since the 1990s, a category thought up by cognitive anthropologists Jean Lave and Etienne Wenger to refer to a group of people who share a craft or a profession (Anderson 2006; Lave et al. 1991).

This concept has since been extended to refer to assemblages of persons (note my wording!) who aim for a common purpose and who are bound together by that purpose. A common purpose can, of course, be ephemeral, although it can also be a powerful rallying cry in our fragmented and dispersed world.

Common purpose also invokes ways of linking affect to practice that people recognize among one another. And from this emerges a fourth aspect of diasporas that Brubaker does not touch on (neither do other diaspora scholars, as far as I am aware): the fact that different diasporic nodes have a commonality of purpose. As several authors in this special issue document, transnational Japanese people's mobilization in response to the triple disaster linked together different overseas groups via social media, which I suggest can be analyzed as the development of a community of practice around affective responses to trauma. This certainly resonates with my ethnographic experience of how the Tongan diaspora operates, namely as a vast network of transnational links that tie together groups of people living across vast distances, through the multi-directional exchange of goods, money, people (grandmothers or small children traveling from one diasporic node to another), ideas, fashions, and, I suggest, affects. Affect is thus embedded in a complex and multi-scalar system in which actions have value in two general senses: materially ("a value") and ethically ("values"). These two understandings of value may bleed onto one another, as when material resources become the expression of ethical stance and vice versa.

What emerges from this discussion of aspects of diasporas that have rarely been foregrounded is the central role that *affect* and the social practices that are related to affect play in the construction of diasporas as communities of practice. Thus affect is the link between inner states and social action, and it is inherently reflexive. It is through affect and the actions that derive from it that we become social and political subjects in response to the people around us. This has motivated some scholars to talk about "economies of affect", breaking down the separation between the materiality of economies and the symbolic nature of emotions, feelings and affects (Zelizer 2013; Narotzky et al. 2014: S4-S16). Centralizing the importance of affect in understanding diasporas helps us understand how the trauma of disaster, which generates particularly strong affects, is linked to the way in which people dispersed around the globe understand themselves, their relationship to each other and their relationship to what many still consider their homeland.

Acknowledgments

I thank Andreas Niehaus and the other organizers of the international conference "Transnational Responses to Catastrophe: Japanese Diaspora Communities and the March 2011 Triple Disaster" for having invited me to present an earlier version of this paper as a keynote address. I am grateful to Christian Tagsold for his comments on the draft. Some passages are revised versions of passages from my article "Communities and Identities: Fraught Categories and Anchoring Resources" (Besnier 2009: 167-171).

Literature

AGAINST, Joseph; CREED, Gerald. 2006. "Reconsidering Community." In: Creed, Gerald (ed.): *The Seductions of Community: Emancipations, Oppressions, Quandaries.* Santa Fe, NM: School of American Research Press, pp. 3–22.

ANDERSON, Benedict. 2006. *Imagined Communities: Reflections on the Origin and Spread of Nationalism.* 2nd ed. London: Verso.

BARTH, Fredrik (ed.). 1969. *Introduction to Ethnic Groups and Boundaries: The Social Organization of Cultural Difference.* Oslo: Universitetsforlaget.

BELLAH, Robert; MADSEN, Richard; SULLIVAN, William M.; SWIDLER, Ann; TIPTON, Steven
M. 2007. *Habits of the Heart: Individualism and Commitment in American Life*, 2nd
ed. Berkeley: U of California P.
BESNIER, Niko. 2009. "Communities and Identities: Fraught Categories and Anchoring
Resources." In: Reyes, Angela; Lo, Adrienne (eds.): *Beyond Yellow English: Toward a
Linguistic Anthropology of Asian Pacific America*. New York: Oxford UP, pp. 167–171.
BESNIER, Niko. 2011. *On the Edge of the Global: Modern Anxieties in a Pacific Island
Nation*. Stanford, CA: Stanford UP.
BOATENG Boatema. 2004. "African Textiles and the Politics of Diasporic Identity-Making."
In: Allman, Jean (ed.). *Fashioning Africa: Power and the Politics of Dress*. Blooming-
ton: Indiana UP, pp. 212–226.
BRUBAKER, Rogers. 2005. "The 'Diaspora' Diaspora." In: *Ethnic and Racial Studies* 28:
pp. 1–19.
BRUBAKER, Rogers; COOPER, Frederick. 2000. "Beyond 'Identity'". In: *Theory and Society*
29: pp. 1–47.
ERIKSON, Erik. 1959. *Identity and the Life Cycle: Selected Papers*. New York: International
Universities Press.
HANNAM, Kevin; SHELLER, Mimi; URRY, John. 2007. "Mobilities, Immobilities and
Moorings." In: Urry, John: *Mobilities 1*. Cambridge: Polity P., pp. 1–22.
JOSEPH, Miranda. 2002. *Against the Romance of Community*. Minneapolis: U of Min-
nesota Press.
LASCH, Christopher. 1979. *The Culture of Narcissism: American Life in an Age of Dimin-
ishing Expectations*. New York: W. W. Norton.
LAVE, Jean; WENGER, Etienne. 1991. *Situated Learning: Legitimate Peripheral Participa-
tion*. Cambridge: Cambridge UP.
LEE, Helen. 2003. *Tongans Overseas: Between Two Shores*. Honolulu: U of Hawai'i P.
MCELHINNY, Bonnie; DAMASCO, Valerie; YEUNG, Shirley; DE OCAMPO, Angela F.; FEBRIA,
Monina; COLLANTES, Christianne; SALONGA, Jason. 2009. "Talk about Luck": Coher-
ence, Contingency, Character, and Class in the Life Stories of Filipino Canadians in
Toronto." In: Reyes, Angela; Lo, Adrienne (eds.): *Beyond Yellow English: Toward a
Linguistic Anthropology of Asian Pacific America*. New York: Oxford UP., pp. 93–110.
NAROTZKY, Susana; BESNIER, Niko. 2014. "Crisis, Value, and Hope: Rethinking the
Economy." In: *Current Anthropology* 55: pp. 4–16.
ORTNER, Sherry. 2005. "Subjectivity and Cultural Critique." In: *Anthropological Theory*
5: pp. 31–52.
SAND, Shlomo. 2009. *The Invention of the Jewish People*. London: Verso.
SMALL, Cathy A. 2011. *Voyages: From Tongan Villages to American Suburbs*, 2nd ed.,
Ithaca, NY: Cornell UP.

Niko Besnier

STOLLER, Paul. 2002. *Money Has No Smell: The Africanization of New York City.* Chicago: U Chicago P.

TAYLOR, Charles. 1989. *Sources of the Self: The Making of Modern Identity.* Cambridge, MA: Harvard UP.

WAYLAND, Sarah. 2004. "Ethnonationalist Networks and Transnational Opportunities: The Sri Lankan Tamil Diaspora." In: *Review of International Studies* 30: pp. 405–426.

ZELIZER, Viviana A. 2013. *Economic Lives: How Culture Shapes the Economy.* Princeton, NJ: Princeton UP.

Mourning for Whom and Why?
3/11 and the Japanese in Düsseldorf, Germany

Christian Tagsold

1 Introduction

The events of March 2011 in Northern Japan were not simply a locally confined disaster but caught the attention of global media immediately. The disaster also crossed regional and national borders in other respects. It did for example strongly remind Japanese in Germany of their *Migrationshintergrund* (immigrant background). This term often used not only in official documents and speeches but also in newspapers started out as neutral denomination in the last decade but has gained a somewhat exclusivist undertone since. Television stations and newspapers were important agents for turning all people of Japanese descent into *true* witnesses of the disaster by assuming that they must have known much more about the ongoing situation and had much stronger feelings of grief because of their background.

In my paper, I want to scrutinize the complex web of assumptions about otherness and othering by using the Japanese in Düsseldorf as an example. Düsseldorf is the hub for Japanese in Germany for various reasons. About one third of Germany's Japanese live here. As a consequence, the diaspora became a focus of attention after 3/11 in the quest of media to establish a rapport with *true* Japanese in order to get the *real* emotions. Helpful in this respect were institutional reactions in Düsseldorf on both the German and the Japanese sides. For the diaspora and citizens of Düsseldorf in general, the city, the state of North Rhine-Westphalia and other institutions staged public events of mourning that then were reported by local and state media and transformed into tangible tokens of the disaster far to the East.

While scrutinizing the interplay between Japanese in Düsseldorf, various institutions, and different levels of media attention, my paper will question whether public events of mourning in this context can become meaningful. I argue that they stay hollow symbols and do not help to foster understanding between the diaspora and other people in moments of deep grief and instead cement feelings of strangeness and alterity; one reason for this is their inherent staging for the media as well as the shared notion of a fundamental cultural gap between Japanese and Germans as well as other people living here.

To show that the Japanese diaspora in Düsseldorf is the best field in Germany to prove my cause and to give more background, I will first take a look on the national level and how the quest of various media channels to find the archetypical Japanese in Germany to relate the disaster far away to the world of the reader or viewer was not bound to Düsseldorf but took place in the whole country. I will then zoom into the Japanese ethnoscape – as defined by Appadurai (1996) – in Düsseldorf and show that the mutual othering that took place after 3/11 is not limited to the disaster, albeit the public events of mourning emphasized this tendency. I will the look at a few examples of mourning and their treatment in media. Finally, I will analyze the outcomes of these rituals on a theoretical level.

2   3/11 in Germany

The triple catastrophe of March 2011 in Northern Japan unfolded quickly in the media worldwide. On March 11, first the news of a massive earthquake in Japan followed by a tsunami were reported everywhere although the full extent of the disaster was not immediately understood. Gradually it became clear that the number of victims would rise to more than ten thousand. During the next days, the looming nuclear disaster in the nuclear power plants of Fukushima added to the horror felt even far away. In Germany, it was especially the nuclear aspect of the triple catastrophe that began to occupy the newspapers and television stations. The general discussion about the perils of nuclear energy that had cooled down somewhat two-and-a-half decades after the Chernobyl disaster became very heated once again and was the main focus of media attention in the weeks following the initial shock (Russ-Mohl 2012).

Since Chernobyl, politicians had heatedly debated the use of nuclear energy. A few years earlier, the ruling coalition of Social Democrats and the Green Party had decided to abandon nuclear power within the next decades, a decision cancelled by the ensuing conservative-liberal government. However, after 3/11 the tide turned again, and Chancellor Angela Merkel came under strong pressure to reinstate the initial decision to end nuclear power.

Within this context, national leading newspapers and television channels, including those regionally bound, tried to explain the disaster in general and the details of the nuclear crisis through explaining its Japanese character, which meant that Japanese scholars and East Asian specialists were contacted for interviews and expertise. Typical questions were why the Japanese had built nuclear power plants in potential danger zones in the first place, why they remained seemingly calm and untouched even in the midst of this major disaster, and finally, why they did not flee the danger of becoming contaminated by radiation immediately. These and similar questions were wrongly based on essentialist notions of a Japanese national character in the first place and thus hard to answer quickly. Answers would become much more complex than most of the journalists wanted them to be.

Many stories of media contacts have been discussed intensively between East Asian specialists in Germany. Within a group of people of the mailing list of East Asian scholars, a heated debate evolved around the story of an interview request by *Der Spiegel*, the leading German weekly political journal, with many of the involved taking sides. The general consensus was that most reporters looked for answers that would portrait Japanese as stoic samurais dealing heroically with a natural disaster.

The media sought out to interview Japanese living in Germany in order to lend a human face to news coverage and if possible to gain additional insights into what the ongoing disaster meant for Japanese. Since there are about 30,000 Japanese living in Germany, it was not too difficult to find enough willing interview partners to fill the void. Yet media stations sometimes wanted more than simply interviews and looked for footage with a specific Japanese touch. A colleague in Munich reported being contacted by the Bavarian State Television, which sought to take pictures of Japanese living in the region mourning at a local Buddhist temple. However, only a few hundred Japanese live in the region even though Bavaria is the economic powerhouse of Germany and Munich the third-largest city of the country. My colleague had to tell the producer of the program that no Buddhist temple exists for this small diaspora and that even if there was a temple, it would be highly unlikely that any Japanese would mourn there. In contrast to the assumption of the producer, not all members of the Japanese diaspora are Buddhist. Even if they were Buddhist, many would see no need to mourn at a Buddhist temple but would rather inform themselves about the ongoing disaster on television.

Many journalists thus attempted to generalize and stereotype the Japanese. In the midst of a quickly unfolding massive-scale disaster, it was on the one hand surely tempting to look for quick answers and on the other simply impossible to fully grasp the complex background. Some journalists tried to do so. I personally was contacted by someone who was quite relieved to learn that most depictions of the Japanese were just based on stereotypes. Yet the overall impression of the Japanese, regardless of whether they lived in Japan or in Germany or anywhere else in the world, was that they are highly homogenous and very different than Germans. This othering through media coverage gained credibility because official Japanese institutions in Germany largely backed these arguments. The discourse of othering finds its counterpart in the so-called *nihonjinron*, the theories of "Japanese-ness" in Japan that are upheld officially by Japan and many Japanese because nihonjinron publications are pervasive on the Japanese market.

3   Ideal Foreigners

All in all, the Düsseldorf area drew the most attention from these media attempts because of the many Japanese living there and the established Japanese infrastructure. While other cities like Berlin, Hamburg and especially Frankfurt also have smaller Japanese diasporas, the one in Düsseldorf is the largest and by far the best established and best known in Germany (Glebe 2004; Tagsold 2010). Düsseldorf prides itself on hosting Little Tokyo, a stretch of Japanese bookstores, food shops and hotels near the main station.

In early March 2011, pundit Henryk M. Broder compared the Japanese community in Düsseldorf to Turkish migrants in Germany in his op-ed published in the conservative newspaper *Die Welt*. Broder is a very well-known albeit somewhat controversial author. His criticism of Islamic migrants and their communities is often harsh. However, the op-ed mirrors the official attitude very well and probably the attitude of many citizens in Düsseldorf as well. The Japanese, according to Broder, resemble the Turkish migrants in one regard: both communities form "parallel-societies". This term was coined in the German discussion on migration and integration during the 1990s (Ronneberger 2009; Yildiz 2009). Conservative and liberal politicians have mostly used the term to criticize the unwillingness of migrants to integrate and thus legitimate state programs designed to govern these migrants' assimilation into German society. Broder (2011) argues that the Japanese parallel-society causes no social problems in stark contrast to the Turkish and asks rhetorically:

> Is it because no Japanese has taken legal action to fight for a prayer room in school? Or because no Japanese has refused to arrange beverages in a supermarket because his religion does not allow him to drink them? Or because Japanese are under-represented as serious offenders but over-represented for high-school diplomas?

Broder's arguments mirror the approach of Düsseldorf's public relations strategy to present the Japanese as "ideal foreigners" as I have argued elsewhere (Tagsold 2010). The city prides itself on the Japanese community – not only on its homepage but also in many other contexts. Although the roughly 6,300 Japanese in Düsseldorf (Jäschke 2008) are a minor group among the more than 130,000 citizens with a foreign passport, they receive by far the most attention. Broder's arguments are widely shared by the city. The Japanese are ideal foreigners who demand no rights and "integrate" with nearly no effort expended by the city. Instead official Japanese institutions regularly help to stage events that are ideally suited for the city's marketing strategies because of their exoticism. The biggest of these events is Japan Day, which is celebrated each year in late spring and brings nearly one million visitors to the city.

Broder's op-ed is marred by erratic assumptions. His essentializing approach, which treats the Japanese and Turks as completely homogenous groups, is a blatant misrepresentation of the complex realities of immigration issues in Germany. Additionally, Broder must completely ignore the composition of the two groups in Germany in order to draw his highly polemical conclusions. While

many Turks suffer from racism and social exclusion and are pushed system-
atically into low-paid jobs or dependence on social welfare, many Japanese in
Düsseldorf work for Japanese companies in extremely well-paid white-collar
jobs or are family members. In addition, the majority of Japanese reside in Germany on a fixed-term basis and will return to Japan after a few years. Thus,
integration and fight for civil rights is by far not as problematic as in the Turkish case. The Japanese diaspora indeed seems to live a quiet life without major
troubles in Düsseldorf, just like Broder suggested in March 2011. This refers to
the image drawn by Düsseldorf's marketing. In showcasing the Japanese, the
much larger communities of Turks, Greeks and Italians get less attention, and
their demands and needs often compare unfavorably to those of the Japanese.

However, the makeup of the Japanese diaspora in Düsseldorf has been shifting
in the last two decades. While rotating white collar workers and their families
were indeed the overwhelming majority twenty years ago, lately more Japanese
tend to stay for more than the three-year period, which is the typical length of a
company assignment abroad (Tagsold 2010: 149). The number of Japanese who
have attained the right to an unlimited stay in Germany is a clear indication of
this. In 2010, more than 2,500 Japanese had attained this status (Gaimusho ryōji
seisakuka 2010: 50), which currently is granted after five years of residence. In
addition, some do not intend to return to Japan and thus have different needs
regarding their surroundings compared to the white-collar strata. This shift is
not reflected in most of the official dealings with Japanese in Düsseldorf, and the
local media still presents the Japanese as temporary guests in the city and not as
full citizens. The aftermath of 3/11 in Düsseldorf gives ample evidence of this
understanding.

4   Staging Solidarity

Many NGOs and other types of groups planned events in Düsseldorf in the weeks
following the triple disaster. Most of them were organized to raise money for
the survivors in North Japan. The Japan Club, which is the main organization of
Japanese in Düsseldorf, performed at the main station and held concerts and exhibitions as charity events. Newly formed groups joined well-established NGOs

in the efforts to collect money and show solidarity with the victims and survivors of the triple disaster. However, because these events were small and scattered, they mostly attracted only limited attention.

Three public events stood out in Düsseldorf because of their size and the attention they got both through attendance and coverage in local media. Two events took place on Saturday ten days after the triple disaster: the memorial service at the local Buddhist temple and the candlelight vigil in the center of the city. The official state memorial service was held a week later on Sunday in a peripheral park. They were all staged as public mourning events open to everyone. They did not involve substantial fund raising but were meant as tokens of solidarity with the victims and survivors in North Japan. The three events assembled a few hundred participants each. Because of attendance and media coverage, these three public events serve best to analyze the most salient problems of the public reaction in Düsseldorf to the tragedy in Northern Japan.

The first of the events was the commemoration service at the Buddhist temple. Roughly 250 people assembled there. The temple is a central institution representing Japan in Düsseldorf. It belongs to one of the main Buddhist schools of Japan. Even though the majority of Japanese in Düsseldorf are not members of these schools, the temple acts as a hub for many of them through its cultural events and its kindergarten. Furthermore, it is deeply entangled with the Japanese consulate general.

During the memorial service, the leading Buddhist priest not only performed various sermons and rituals but also read recollections of victims of the catastrophe and thereby gave a voice to those who suffered most. Except for the chanting of sutras, most of the ceremony was conducted in German. The attendance was mainly German. Only a few dozen Japanese were present – a rather unusual ratio of participants for the temple, which normally attracts more Japanese than Germans to its events. In contrast to the low Japanese attendance, media was very present at the service. Three television teams, two belonging to national state stations and one to a local private channel, actively moved around to get good footage. In addition, newspaper journalists from local newspapers waited in front of the temple to interview participants. Media representatives made up more than 10 percent of the attendees.

The candlelight vigil on the evening of the same day in the center of the city was also a focus for the media. However, journalists only took pictures during the first minutes and conducted short interviews with the organizers before rushing back to their editorial departments. Two high-school students had organized this commemoration in order to offer a chance to citizens of Düsseldorf to express their grief in a non-political context, as they told me in an interview. About 500 people followed the call to line up while holding candles in their hands. The mode of the commemoration was a little bit odd because candlelight demonstrations usually are a form of protest against right-wing violence against immigrants that sprang up in the 1990s. Participants whom I interviewed partly had a background of such protests and even had brought along their candles from the old days. Other participants strongly sided with organizer's point of holding an apolitical commemoration.

Finally, the official commemoration event of the state of North Rhine-West-phalia took place one week later. The president of state opened the event with her speech followed by the mayor of Düsseldorf and the Japanese consul general. In addition, the priest of the aforementioned Buddhist temple again conducted a Buddhist ceremony. At the end of the event, a Catholic and a Protestant priest read a joint Christian message before the Buddhist priest struck the bell of the Buddhist temple to conclude the event. In stark contrast to the other two events, a few hundred Japanese took part this time, nearly outnumbering German attendance. The Japanese mostly attended in black mourning suits while the Germans were dressed in usual Sunday afternoon dress. This commemoration had some political undertones. The official speakers at the state memorial service subtly interpreted the events through their lens – the conservative mayor of Düsseldorf tried to avoid the topic of Fukushima while the left-wing state president made it the main topic of her speech.

The event had another unsettling layer of meaning. Typically at such events in Germany not only political leaders express their grief and solidarity but also religious leaders are invited to console participants. Thus, a Buddhist priest, a

Catholic priests and a Protestant pastor took part and conducted religious ceremonies. Nevertheless, the Buddhist ceremony did obviously puzzle German participants. The Buddhist priest conducted his part standing with his back to participants and chanting mostly sutras. No explanation accompanied his actions. Most Japanese closed their eyes and took an appropriate pose, but many Germans looked around bewildered. In contrast, the Catholic and Protestant representatives spoke to the crowd explaining their religious interpretation of the catastrophe.

However, local media coverage created new interpretations contradicting the impressions from the spot. According to the local press, the Buddhist ceremony at the temple and even the candlelight vigil seemed to be deeply Japanese expressions of grief. Both public events were depicted and described as being dominated by Japanese participants. This contrasted with the real attendance at the events. For example, the leading local newspaper on the next day depicted the few Japanese at the candlelight vigil in about half the pictures on their homepage coverage of the events. In contrast, local media only cursorily mentioned the state commemoration one week later and even less so the many hundred Japanese present there.

A reason for this type of local media coverage surely can be explained by the visual qualities of the public events. The Buddhist temple and the candlelight vigil lend themselves perfectly for stunning photos. In addition, both settings allowed for instant othering by highlighting the Japanese attendance. This not only made the events more exotic but also legitimized them as authentic expressions of those affected most by the triple catastrophe. The official state commemoration had no such qualities. Most Japanese attended in black suits that were apt to express their feelings but not very exotic.

However, it would be too easy to speak of medial misrepresentation through media coverage in the case of the memorial events. All three public events were not simply staged for those attending but had been organized with local media coverage in mind. The state commemoration offered a platform for cameras and journalists, which even hindered the crowd to see the stage fully. At the Buddhist temple, media had been given full access and had been allowed to park their vans on the grounds next to the temple itself. Finally, the pupils organizing the candlelight vigil had learned their lessons quickly and were freely giving interviews to the journalists. Thus, media was part of the commemoration from the very outset in all three cases.

5   Empty Rituals

Once Clifford Geertz (1973: 448) asked what people "tell themselves about themselves" in rituals. I have deliberately avoided the term "ritual" for the three cases and instead used Don Handelman's (1998) alternative "public events". But setting this differentiation aside, which I will take up again in a moment, Geertz's questions clearly needs a much more complex answer in the case of Düsseldorf. The public events are not set in one cultural set but at least two – the Japanese one and that of Düsseldorf itself. As a consequence it is not so clear who "themselves" denotes. In addition, these events did not simply gain meaning through being staged on the spot and only for those being present. In contrast, local media coverage lent them a deeper meaning, thus broadening the notion of "themselves" in yet another dimension. As a consequence, the commemoration of the catastrophe in Düsseldorf was extremely multilayered and multifaceted. What is true for Düsseldorf also holds true for Germany in general as the examples in chapter 2 prove; however, each and every public event did not expand its meaning along the lines I will analyze now.

Handelman (1998) has divided public events into two types. One type of these events mirrors, the other models. Handelman, thereby, was able to analyze what anthropologists have defined as rituals in a new way. Some public events change the world through modeling them. They ingrain a new vision of the world in their design and apply it to this very lived-in world. Other public events simply reflect what is out there already and amplify it. Bureaucracies in modern states are often the sponsors of public events as mirrors. They propagate their taxonomies through staging them. But mirrors are not only set up by the state or institutions belonging to it. Many events in the modern world are ruled by official taxonomies.

Handelman helps to understand the three public events in Düsseldorf and the role media did play. The case seems most clear for the state commemoration, in which bureaucracy seemed to control the event and its outcomes fully. The value of international solidarity was staged by the state of NRW, the city of Düsseldorf, and the Japanese consulate general. In doing so, a clear divide between Germans and Japanese was confirmed – the former showing their solidarity with the latter but not overcoming what separates both groups. In that sense, the public event acted as a reinforcement of the notion of parallel societies.

The other two events were less designed to differentiate between Japanese and non-Japanese. They addressed both groups equally. Especially the candlelight vigil seemed to work on a strong symbolic unification of all citizens. However, nearly no Japanese attended because the symbolism of the event probably was lost to them. They had not experienced the huge candlelight vigils two decades earlier that *did* foster simultaneously a strong mood of integration and multiculturalism. The organizers had not been able to convey this to the Japanese living in Düsseldorf.

Local media worked against the possibility of integration through shared mourning. By singling out the Japanese for sake of making pictures of the event more exotic, the newspapers fed the trend of othering again. The design of the candlelight vigil helped them strongly to achieve this effect. The same was true for the whole setting of the Buddhist temple. In contrast, the official state mourning did not offer many exotic impressions but the usual political staging. In addition, it came too late to leave much of an impression in the media even though it had the largest group of Japanese of all events and also the most prominent mourners.

Handelman (1998: XXXVIII) has pointed out that a strong emphasis of visual qualities is a typical trait for public events that mirror. This is very much true for the two events at the Buddhist temple and the candlelight vigil. They did answer many questions through visualizing solidarity explicitly and otherness and exoticism implicitly. Local media coverage amplified these visual qualities in both cases. The mourning at the Buddhist temple even made it into national prime time news, but the few seconds shown on TV were only meant as a glimpse into exotic Japan in Germany. The colorful and somewhat mystical atmosphere of the temple added to the overall strangeness of the ongoing crisis in Japan after 3/11.

6 Conclusion

On March 11, 2012, many remembered the horrible triple catastrophe. The nuclear wreckage of Fukushima clearly was the focus of most of the media coverage because this part of the catastrophe has not yet been resolved and will not be for many decades. In contrast, the misery that the earthquake and most of all the tsunami had inflicted on people in Northern Japan was, if not forgotten, at least much less the focus of public memory in Germany.

What is somehow telling is that the public events of 2011 that were meant to prove solidarity with Japan had evaded the public memory. They had left nearly no imprint on the quality of relations between the Japanese in Düsseldorf and other citizens. Neither the Internet forum Dusselnet, which gathers voices from Japanese in Düsseldorf, nor the consulate general, during their year-end party celebrating the birthday of the emperor, mention any of these events. The attention of the consulate general had shifted to display Japan in general and specifically Northern Japan as healthy overall with much to offer for export. Reestablishing economical viable export relations was paramount at the end of 2011 at the consulates general party.

This is yet another proof of how empty signs of solidarity were in shaping the image of Japanese living in Düsseldorf. After all, the events were important for momentarily adding a local touch to world news but not important in their own right. This is a conclusion that certainly holds true for Japanese in Germany in general. These Japanese did not gain a heightened interest and even less so were seen as members of a diaspora. In contrast, public events after 3/11 and especially their representation in local media othered Japanese living in Düsseldorf. Instead of turning them into co-citizens, they idealized Japan.

On the personal level, some stories of the Japanese reflect the problems of the public events and their representations while others point in a different direction. One writer on Dusselnet complained in her entry that she felt as though Germans scolded her, for example, in the supermarket, for the nuclear catastrophe of Fukushima. As a Japanese, she was made responsible in her eyes for a misguided nuclear policy with too low standards of security. However, other Japanese reported that neighbors and friends tried to console them and offered help.

Literature

APPADURAI, Arjun. 1996. *Modernity at Large: Cultural Dimensions of Globalization.* Minneapolis: U of Minnesota Press.

BRODER, Henryk M. 2011. "Warum japanische Parallelgesellschaften keinen stören" [Why Japanese Parallel Societies do not Disturb Anyone]. In: *Die Welt*, March 3, 2011, http://www.welt.de/debatte/henryk-m-broder/article12688016/Warum-japanische-Parallelgesellschaften-keinen-stoeren.html. (accessed November 25, 2015)

GAIMUSHO ryōji seisakuka. 2010. "Kaigaizairyūhōjin chōsa tōkei"[Statistical Results for Japanese Living Abroad]. http://www.mofa.go.jp/mofaj/toko/tokei/hojin/10/pdfs/1.pdf. (accessed November 15, 2012)

GEERTZ, Clifford. 1973. *Interpretation of Cultures: Selected Essays.* New York: Basic Books.

GLEBE, Günther; MONTAG, Birgit. 2004. "Düsseldorf: Nippons Hauptstadt am Rhein" [Düsseldorf: Nippon's Capital on the River Rhein]. In: Frater, Harald; Glebe, Günther; Looz-Corswarem, Clemens von (eds.): *Der Düsseldorf-Atlas: Geschichte und Gegenwart der Landeshauptstadt im Kartenbild.* Cologne: Emons, pp. 74–77.

HANDELMAN, Don. 1998. *Models and Mirrors: Towards an Anthropology of Public Events.* New York: Berghahn Books.

JÄSCHKE, Ruth. 2008. "Japaner in Düsseldorf" [Japanese in Düsseldorf]; http://www.dus.emb-japan.go.jp/profile/deutsch/kulturbuero/japaner_in_ddorf_2008-06.htm. (accessed May 4, 2014)

RONNEBERGER, Klaus, TSIANOS, Vassilis (2009). "Panische Räume: Das Ghetto und die Parallelgesellschaft." [Panic Spaces: The Ghetto and the Parallel Society] In: Hess, Sabine; Binder, Jana; Moser, Johannes (eds.): *No Integration!: Kulturwissenschaftliche Beiträge zur Integrationsdebatte in Europa.* Bielefeld: Transcript, pp. 137–152.

RUSS-MOHL, Stephan.: *"Haben Journalisten die Energiewende herbeigeschrieben?"* [Have Journalists Written the Energy Transition Into Being?]. NZZ, 25.09.2012; http://www.nzz.ch/aktuell/feuilleton/uebersicht/gefaerbte-informationen-ueber-fukushima-1.17639366. (accessed November 15, 2012)

Christian Tagsold

TAGSOLD, Christian. 2010. "Establishing the Ideal Foreigner: Representations of the Japanese Community in Düsseldorf, Germany." In: *Encounters.* 3 (1): pp. 143–166.

YILDIZ, Erol. 2009."Was heißt hier Parallelgesellschaft": Von der hegemonialen Normalität zu den Niederungen des Alltags." [What's that supposed to Mean: Parallel Society: From Hegemonial Normality to the Banalities of Everyday Life] In: Hess, Sabine; Binder, Jana, Moser, Johannes (eds.): *No Integration!: Kulturwissenschaftliche Beiträge zur Integrationsdebatte in Europa.* Bielefeld: Transcript, pp. 153–70.

"Even if it is Just a Little Help for the Victims from the Distant Belgium": Japanese Nationals in Belgium and the 3/11 Triple Disaster

Andreas Niehaus and Tine Walravens

1 Introduction

Anthropologists have argued that disasters "mobilize forces of cultural change" and that the potential fields of studies vary because disasters affect every aspect of human life, be it social, economic, environmental, political or even biological (Hoffman et al. 2002: 3–22). The theoretical approaches on disaster, however, take as their subject of study members of societies and cultures who were directly struck by disaster and limit their research spatially to the inflicted areas. Nevertheless, members of a given diaspora will also be deeply affected by catastrophe in their homeland. Rogers Brubaker (2005: 12) remarks that "as a category of practice, 'diaspora' is used to make claims, to articulate projects, to formulate expectations, to mobilize energies, to appeal to loyalties". It is, we will argue, in times of crisis and trauma that these practices construct, increase and intensify an awareness of community and evoke certain reactions generated by emotions, feelings and affects. When Japan was struck by the 3/11 Triple Disaster in 2011, Japanese nationals living in Belgium took a diasporic stance and immediately showed their commitment and loyalty by organizing charity events and moral support activities.[1] Edith Turner (2012: 76) coined the term "communitas of

---

[1] Figures obtained from the Japanese Embassy in Brussels show that all the overseas establishments of Japan (meaning, for example, all embassies and consulates, JETRO offices abroad) received J¥ 8.9 billion (= € 88.5 million) as of February 2012. The total amount which the Belgian Red Cross transferred to its Japanese counterparts is € 1,061,129 or J¥ 116,563,464 of which € 666,129 was from the Flemish Red Cross and € 395,000 from the Croix-Rouge de Belgique, Communauté francophone. Of that total amount, € 22,000 came from gifts received by the Japanese Embassy in Belgium and € 315,000 from gifts received by the BJA, the Belgium-Japan Association and Chamber of Commerce. These figures do not include donations and money from support actions organized by the Belgian community in Japan, nor donations transferred directly to the Japanese Red Cross or other organizations in Japan. Moreover, it seemed impossible to analyse what percentage of this money that has been donated by Belgians, Japanese individuals or companies in Belgium. See also Interview 1, May 22, 2012.

disaster" in order to describe the feeling of togetherness that will result from a shared experience of a trauma. The parameter "shared experience", however, can be applied to different groups, creating different communitas as well as degrees of "sharedness" and separation. With regard to the Tōhoku Triple Disaster, "shared experience" firstly refers to the direct victims in Tōhoku and secondly to their families not living in the affected areas.[2] However, "shared experience" can also refer to the idea of the Japanese nation, in contrast to other nations that were spared from the catastrophe. A further level of sharedness existed between the Japanese nationals living abroad at the time of the disaster and the Japanese nationals living in the home country. Although the group of Japanese living abroad long term or permanently still are and still feel Japanese, they will often be considered not Japanese "enough" by the Japanese living in Japan and by that denied "ownership" of the catastrophe. The disaster that struck Northern Japan thus creates a complex field where questions of identity, identification and belonging are tested and negotiated.

The disaster created diasporic communities that gave gifts with the purpose of supporting the homeland both materially and morally. Niko Besnier, referring to Joel Robbins, argues that these gifts also strengthen allegiance to the homeland: "intrinsic to giving and accepting is the mutual recognition of the other party, and thus the basis of self-conscious selfhood, which places the gift right at the center of a moral order. Remittances, then, bind the participants in a common moral order of mutual recognition. And so do that particular kind of remittances, gifts provided to help relief efforts in the wake of disaster" (Besnier 2009: 71–80). By linking remittances with the "crisis of return", we could argue that gift-giving will serve to show the solidarity of an individual as well as a group with the homeland, thus strengthening the feeling of belonging on both sides and securing a smooth return to the homeland for the short-term expatriates. In this context, we should also mention that donating and showing support manifested as a practice of communality; thus, it also created a sense of belonging (or exclusion) within the group of Japanese nationals living in Belgium.

---

[2] The different levels of shared experience also applies to direct victims of Tōhoku because some will have lost family members or friends, while others will "only" have lost their property or material possessions.

Based on fifteen in-depth interviews with Japanese nationals living in Belgium conducted between March and December 2012,[3] this article will analyze how far their response as a social practice differed in relation to their residential status and to what extent they negotiated questions of identity, communality and group consciousness by engaging in support projects. Diasporic communities can generally be divided into short-term and long-term residents. For this study, we chose Brussels (the capital of Belgium) and Ghent (in the Flemish region) because the Japanese diasporic population in the former is characterized by a large number of short-term rotating individuals with a low degree of integration, while the Japanese nationals living in the latter are mainly permanent residents that are generally well integrated into the host community. In Ghent, we interviewed one short-term and five long-term Japanese residents. In Brussels, we conducted a total of nine interviews, seven of which were with Japanese short-term residents in Brussels. Additionally, we also interviewed a Japanese official at the Embassy of Japan and a Belgian representative of the Belgium Japan Association (BJA) Chamber of Commerce. We based our main approach on the assumption that diasporic communities are indeed characterized by practice. However, before focusing on how practice (re)created community following the triple disaster, it is necessary to map the ethnoscape of Japanese nationals in Belgium, based on the available official statistical data as well as the community's infrastructure.

2   Mapping the Ethnoscape of Japanese Nationals in Belgium

The Kingdom of Belgium, with a population of 9,832,010 (in 2011), is divided into three parts. The division is based on language, which, amongst other issues, fuels political and social conflict. Belgium has three official languages: Dutch, French and German. The Dutch speaking community is situated in the economically strong northern region of Flanders and the French speaking population in the Southern region of Wallonia. A minority of German-speaking Belgians lives

---

[3]   The interviews focused on questions concerning personal life, daily habits, interpersonal relations and network dynamics, organization of and participation in charity events, impact of the disaster on identity and feelings of belonging, media coverage.

in the Eastern part of Wallonia. According to the most recent statistical data provided by the EUROSTAT population census, a total of 4,458 Japanese nationals
were living in Belgium on January 1, 2011, of which 2,020 were male and 2,438
female.[4]

The majority of Japanese nationals who come to Belgium do so for remunerative reasons or family reunification with a Japanese national and will only stay
short-term.[5] However, Belgium also has a registered group of 901 long-term
residents with Japanese nationality.[6] Over the last three years, the number of Japanese nationals living in Belgium decreased due to the growing economic crisis
facing Japan, which meant that companies reduced their overseas staff members,
closed overseas offices or decided not to open new offices. The data received
from the statistics show the Japanese expatriates in Belgium as comparable with
other communities in Europe, for example, in Germany and the Netherlands.
On the one hand, there are rotating, short-term residents of Japanese, primarily
male, employees and their families working for transnational companies or for
governmental institutions and whose residence is concentrated in urban centers
such as Düsseldorf and Amsterdam. On the other hand, we find a primarily female group of long-term residents married to Belgians.[7] Although its members

---

[4] According to ADSEI (2012), there was a total population of 4,458 Japanese in 2011. As of January
    1, 2010, a total of 4,543 Japanese were registered in Belgium. For 2010, see also ADSE (2011).
[5] The first permits issued for family reasons and remunerative reasons are interrelated; from a total
    of 512 first permits for family reasons, 373 were issued on the basis of "person joining a non-EU
    citizen". In these cases, Japanese employees are joined by spouses, partners and children. How-
    ever, it is interesting to note that the number of first permits for persons joining an EU-citizen
    more than doubled between 2010 (58) and 2011 (139), whereas the persons joining non-EU cit-
    izens decreased in absolute numbers from 412 in 2010. See Eurostat; Centrum voor gelijkheid
    van kansen en voor racismebestrijding (2011). First permits issued for remunerative activities ac-
    counted for 334 permits in 2011, with the majority issued for a period less than a year (303). Only
    62 permits were issued for educational purposes, of which 58 were for studies, and 54 received
    a permit for less than a year Eurostat (2012). The number of Japanese students coming to Belgium
    dropped from 91 in 2008 to 58 in 2011.
[6] In 2011, a total 901 Japanese citizens were registered as long-term residents in Belgium of which
    72.6 percent (655) were female; (Eurostat).
[7] When gender is taken into consideration, it shows that the first permit visa for remunerative ac-
    tivities in 2011 were mainly issued to male Japanese, accounting for 89.4 percent (262), with only
    31 permits issued to female Japanese. As can be expected, the ratio is reversed in the case of
    first permits for family reasons (Eurostat). See also Centrum voor gelijkheid van kansen en voor
    racismebestrijding (2012).

are primarily middle-class and residing in Belgium due to corporate rotation, the transient community in Belgium is quite diverse and will, as became evident from the interviews, generate formal and informal, mobile subgroups, based on occupation (company), place of residence, age of the children, gender or leisure interests.[8]

The highest concentration of Japanese citizens (2,987 in 2011) can be found in the county of Brussels.[9] As Brussels is the capital of Belgium, home to the NATO headquarter and primary seat of the EU, it attracts a great number of Japanese companies and lobby organizations.[10] It also houses Japanese governmental and semi-governmental organizations, including the Japanese Embassy to the Kingdom of Belgium, The Mission of Japan to the EU, EU-Japan Centre for Industrial Cooperation, Japan External Trade Organization Office (JETRO) and the Belgium-Japan Association and Chamber of Commerce (BJA, founded in 1991). Brussels houses a large international corporate and political expatriate community, whose members are well paid. They live in the more exclusive parts of the city and their social contacts often stay within the limits of this international expatriate community. Also, the Japanese living in Brussels as rotating, short-term residents are not well integrated into the host city but constitute a parallel society. However, the Japanese residents receive little to no attention from the host country. It seems that there is no direct need for the Japanese to integrate into Brussels' society.[11] Brussels is actually situated in the Flemish part of the country but is mainly French speaking. Within the circles of the international corporate and diplomat community of Brussels, English and French

---

[8] This pattern can also be observed in other European cities with a high concentration of Japanese, rotating short-term residents. See White (2003) and Befu (2001).

[9] Reflecting the economical significance of the counties, 1,230 Japanese citizens live in Flanders (excluding Brussels) and 240 in Wallonia. See ADSEI (2011). The concentration of 3129 (2010) Japanese citizens living in Brussels makes the economic disparity even more significant when we take into consideration that 551 Japanese nationals are living in the province of Flemish-Brabant, which surrounds Brussels, and in the neighboring province of Walloon Brabant (91), which results in a percentage of 81.4 percent Japanese nationals living in the close vicinity of the capital.

[10] According to the Belgium-Japan Association, Chamber of Commerce (2011: editorial), there are about 220 Japanese companies registered in Belgium. Of these companies, 60 percent are situated in the county of Brussels; Interview 9, June 4, 2012.

[11] This has also been argued by Christian Tagsold (2011: 160).

are the languages most commonly spoken, whereas Dutch plays no significant role. Accordingly, there is no practical reason for short-term Japanese residents to learn Dutch.[12]

It is generally agreed upon that short-term residents tend to reside in close proximity to each other, thus creating a sense of home in the host country. This is also true for the Japanese in the county of Brussels, where the majority of Japanese nationals in Belgium reside. When taking a look at the statistical data available concerning the Japanese population in Brussels, we see that in 2011 a total of 2,987 Japanese nationals were living in this county. Among those living in the county of Brussels, 1,991 Japanese residents are concentrated in only three municipalities from a total of nineteen municipalities in the county; these three municipalities are St. Lambrechts Woluwe, St. Pieters Woluwe and Oudergem (ADSEI 2011). If these data are cross-referenced with the prices of the real estate market in Brussels, we see that St. Lambrechts Woluwe and St. Pieters Woluwe have the highest rent for apartments in Brussels and that prices in Oudergem are also generally above average. Members of short-term rotating diasporas are aware of the fact that they will only stay temporarily in the host country and are thus less inclined to socialize with members of the host country or to adapt their life style. As Katarzyna Cwiertka argues concerning short-term employees in the Netherlands:

> The "employees" follow the logistic decisions of their employers with little influence on their destinations. This, along with the temporary character of their residence, is the most important factor responsible for the formation of suspicious feelings towards foreign culture and a tendency to recreate a Japanese lifestyle outside Japan in order to leave their identity as untouched as possible. (Cwiertka 2002: 148)

However, Merry White also convincingly linked the creation and maintenance of a "Japanese" environment to the question of return, ascertaining that a Japanese life-style will make the re-entry into the Japanese culture easier: "Re-entry raises

---

[12] However, even within the circles of highly integrated, long-term Japanese residents living in the Flemish province, English is generally the lingua franca because the partners often do not speak Japanese; Interview 12, December 10, 2012.

questions of identity that can be silenced only by strict conformity and virtual denial of the foreign experience." (White 1992: 106)[13]

Japanese short-term residents therefore create a sort of cultural island, an "ecology" that mimics the homeland, where members of the community can live a life according to their standards, needs and interests.[14] Furthermore, the hierarchical group structures and social activities of the homeland are transferred and practiced in Belgium. The infrastructure in Brussels, mainly maintained by long-term Japanese residents, includes Japanese supermarkets, Japanese restaurants, Japanese recreational clubs, a Buddhist temple, a bulletin (*Petits Pois*, Jap. *Puchi-Powa*) and the Nihonjinkai (Association of the Japanese), a purely Japanese organization that promotes exchange between Belgium and Japan and supports the Japanese School of Brussels (JSB), which it founded in 1973.[15] This private Japanese school provides an education that conforms to the Japanese educational system on weekdays and includes Japanese language education on Saturday morning (*hoshūkō*).[16] We see that mostly children from the short-term community attend the Japanese curriculum on weekdays, while Japanese staying permanently tend to enroll their children in the Belgian educational system and additionally send them to the Japanese school on Saturday mornings if they live in Brussels or the vicinity.

Because Japanese long-term residents in Brussels are often involved in providing logistic support and knowledge to short-term residents or work for Japanese companies, there is interaction between the two groups on the professional

---

[13] Interviewed short-term resident from Brussels commented on the decreasing numbers of applications from Japanese teachers wanting to go abroad. The interviewee gave the following reasons: low income, missing the development in Japan, and easier life in Japan; Interview 15, October 24, 2012.

[14] The term "ecology" here refers to an infrastructure providing stability, providing a "home abroad", a parallel society, a foreign community that produced its own environmental bubble (Tagsold 2011: 147).

[15] The bulletin, or information sheet (*jōhō-shi*), *Petits Pois* was first issued in 1992 by Japanese housewives and since 1993 has been registered as an association without lucrative purpose. Today, it is published at the beginning of every month (with the exception of August), eleven times per year and has a circulation of 2,000. Furthermore, fundraising and support events are advertised through this bulletin.

[16] The school has about 350 primary and junior high students (Japanese School of Brussels).

level. However, the social worlds and private networks of both groups remain separated, as one interviewed short-term male resident in Brussels also stated.[17]

Belgium is a rather small country, approximately 30,500km^2; however the Japanese nationals living outside the direct vicinity of Brussels, such as the Japanese in Ghent, have hardly any contact with the Japanese nationals living in Brussels, Japanese organizations or even cultural activities that can be found in the capital. This is also true for the Japanese living in the city of Ghent. Ghent has a population of approximately 250,000; the city is situated in the Flemish region about 50 kilometers northwest of Brussels. The number of Japanese nationals registered in Ghent on December 31, 2011 was 62 (Gent City 2012: 66). Against the general overall trend seen in Belgium, the number of Japanese residing in Ghent is actually increasing. The Japanese expatriates in Ghent are dominated by female Japanese who are married to Belgian nationals. Most of those women are working outside of the household, in contrast to the professional housewife of the short-term Japanese community. Through their partners as well as their work environment, they are generally well integrated into the Flemish community, although the interviews suggest that social contacts are mainly with other Japanese or expatriates from other nations. Ghent also lacks the infrastructure for maintaining a Japanese lifestyle, including supermarkets to obtain ingredients for Japanese meals. There are a number of Japanese restaurants located in Ghent, but they depend on non-Japanese customers and are owned by non-Japanese who cash in on the recent sushi hype.

The children of the Japanese long-term residents living in Ghent are enrolled in the Belgian school system and on Saturdays additionally attend the Japanese school in Lille or, to a lesser degree, the Japanese school of/in Brussels. The choice for Lille, which is located just across the border to France, is firstly economically motivated, because the fees there are considerably lower. A second factor in Lille's popularity that was put forward by one of the interviewees is the "Japaneseness" of the school in Brussels:

> Their [school in Lille] target is primarily half-blood children. But, that is my opinion, the quality of education in Brussels is also very good, but they try to

---

[17] Interview 15, October 24, 2012.

do everything as it is done in Japan. The school in Lille is more for children that have another language [than Japanese] as mother tongue.[18]

Whereas the short-term residents choose an education based on the Japanese system in order to make the reintegration of their children into the Japanese school system easier, "doing it the Japanese way" is considered to be a disadvantage for the children of long-term residents because they are used to the Belgian system and will not need to adapt to the Japanese school system in the future.

In summary, we have observed that there are differences between short-term and long-term Japanese residents living in Belgium that range from life style, housing, schooling of children to social life and integration into the host country. In the following chapters, we will show that the aforementioned differences are parameters that have an influence on creating communality, belonging and identity through the practice of supporting the homeland.

3   Supporting Home: Practicing Diaspora in Brussels and Ghent

3.1   Institutionalized Support: Japanese Short-Term Residents in Brussels

Examining the fundraising events and support activities from the rotating short-term community in Brussels, we see that the short-term residents have a low level of participation in the events organized outside the community and in the organization of support actions. If there were actions organized by the short-term community, they were aimed at and directed towards the community: Japanese supporting Japanese.

Institutions such as the embassy of Japan and the Nihonjinkai, an association for and by Japanese living in Belgium, did not organize any support actions. This tendency could be explained by the fact that the Nihonjinkai considers itself as receiving partner.[19] The same is true for the embassy – it did not contact Japanese citizens actively and none of the respondents from the interviews had contacted it. A representative in Brussels explained that the embassy's limited role towards charity actions is due to the Japanese legal framework.[20] Constrained by

---

[18] Interview 12, December 10, 2012.
[19] Interview 9, June 4, 2012.
[20] Interview 1, May 22, 2012.

this framework, the embassy can neither take up a responsible role nor organize charity events by themselves; they are limited to supporting fundraising activities indirectly, for example through patronage. As such, the embassy was a patron of six support events. The embassy acted as an intermediary for the donations and tried to assist by disseminating information on several levels, such as the website and during cultural activities. The embassy also functioned as the official representative of the government and seemed to concentrate on restoring people's trust in the government and economy, thus limiting the economic damage. At an event attended by more than 800 invited guests and held in the Hilton Hotel in Brussels one year after the Triple Disaster, Ambassador Yokota Jun, speaking on behalf of the Nihonjinkai and the Ambassador of Japan to the European Union, Shiojiri Kojiro, openly addressed politicians and policy makers, asking them to lift the import restrictions and to support the Japanese economy.[21] The event also included a minute of silence and promotional wine tasting. Also, JETRO focused worldwide on building up foreign investors' trust in the Japanese economy in general and in the stricken areas by organizing conferences and briefings for companies and business organizations. JETRO Brussels, for example, jointly with the Mission of Japan to the European Union, held a briefing for European companies and business organizations to explain countermeasures taken in response to the Great East Japan Earthquake (The Mission of Japan to the European Union 2011).

The business community in Brussels also engaged in activities, for example, the "HOPE" project of the Japanese car manufacturers in Belgium (Isuzu, Mazda, Nissan, Subaru, Suzuki and Toyota), through which these companies transferred € 20 per car sold between March 11 and April 15, 2011 to the Japanese Red Cross

---

[21] Ambassador Yokota Jun also expressed his regret about EU import restrictions on Japanese products during a memorial service at the Japanese Garden in the city of Hasselt on November 3, 2012, which was attended by about one hundred Japanese and Belgians. As part of the ceremony, Yokota Jun together with the president of the Belgium-Japan Association planted a cherry tree; a plaque donated by the Belgium-Japan Association was placed. The event was framed by the female Japanese choir of Brussels and a group playing Japanese drums. The photograph that was placed above the Internet article in the Belgian newspaper *De Morgen* actually showed Asian faces with white face masks and candles. The photograph was not of the ceremony in Hasselt but of a "memorial demonstration" in Brussels mainly attended by anti-nuclear energy supporters.

Society. A key organization in Belgian and Japanese business and cultural relations is certainly the Belgium-Japan Association, Chamber of Commerce (BJA). The BJA, with a primarily corporate-oriented membership, supported activities (e. g. charity concert by Seikyo Kim, May 13, 2011) and published a call for donations as early as March 14, 2011 (Belgium-Japan Association: 2011: 2). By March 30, thus within a mere fifteen days of the earthquake, they collected € 263,138; this sum was then transferred to the Red Cross (Belgium-Japan Association: 1).

These events and briefings displayed a "male" official face, which stands in contrast to the "female" face of grassroots-level activities organized for the long-term Japanese network in Ghent (see below), where one long-term female resident remarked: "Japanese men organizing Zero."[22] In the case of the short-term resident community in Brussels, it was not individuals which functioned as hubs, it was the organizations and companies. Japanese organizations and companies located in Brussels engaged in calls for financial support, donated money, and logistically supported or participated in events, but generally did not organize events by themselves. When events were organized, they stayed within the community in terms of both population and place. The Japanese School of Brussels can serve as an example; the school was contacted to participate in and promote different activities.[23] The Japanese School of Brussels then decided to engage primarily in events that were somehow connected to children and education, for example with the European School of Brussels and Dyslexia International. They forwarded messages of support from different international schools in Brussels to Japan. The school itself organized a "writing characters of support" event and collected money as well, but these activities stayed within the boundaries of the school.[24] One of the main charity events organized in Brussels was certainly the Japanese charity flea market on April 3, 2011 at the Notre Dame Church in Stockel, Sint-Pieters-Woluwe, a neighborhood with a high density of Japanese

---

[22] Interview 11, November 7, 2012.

[23] However, requests to use the school grounds for fundraising activities were denied.

[24] Interview 14, October 24, 2012. Interview 15, October 24, 2012. The money collected by the Japanese school was partially donated to a primary school in Miyagi through contacts of the former school director.

expatriates. The initiative of the flea market came from a female Japanese long-term resident in Brussels, who had also experienced the Great Hanshin Earthquake in 1995.[25] The flea market was officially coordinated on a volunteer basis by Misc Netto, a website organized by Japanese female volunteers living in Brussels, and the web company Beru tsū. Both support the Japanese short-term community in Belgium by organizing *"sayonara* sales" or by providing useful information on life and living in Belgium.[26] Beru tsū (2011a), with 35,000 visitors per month on their website, is also well connected with social media and hosts a staff blog for members, distributes a "mail magazine" and created a webpage devoted to the triple disaster. The service offered by these websites is crucial for the Japanese expatriate community in Belgium, and the organizers were able to activate a broad network which attracted, according to the organizers, more than 3,000 participants and resulted in € 25,614 from sales as well as donations. The funds were transferred to the embassy, which then donated the money to the Japanese Red Cross. The organizers also reached out to the Japanese School in Brussels,[27] and information about the fundraising events was spread through the short-term community's information bulletin (*minikomi*) *Petits-Pois* (*Puchi Powa*). Thus, the organizers could rely on a well-established institutionalized and structural network. The volunteering participants were generally short-term Japanese residents in Brussels. The flea market also included a Japanese café, chocolate and cake shop, a shiatsu and origami workshop, Japanese drums demonstration and a concert by Japanese musicians, thus creating a "Japanese" atmosphere. Taking into account the special situation of Brussels, flyers were printed in four languages (Japanese, French, Dutch and English) and distributed mainly in the Japanese community as well as in cultural and corporate circles with a strong connection to Japan. The Japanese flyer, in contrast to the French, Dutch and English versions, explicitly made clear that the goods sold were donated following a call for donations and that all participants in the event, including the staff members of the organizing companies, were volunteers and that accordingly 100

---

[25] Interview 15, October 24, 2012.
[26] The company name Beru tsū can be translated as "Belgium Connoisseur". "Beru" can also mean "bells", which are represented in the logo.
[27] Interview 15, October 24, 2012.

percent of the earned and donated money would go to the victims. Reading the Japanese language website of Beru tsū, it becomes clear that the charity bazaar is seen as an activity of a group of Japanese nationals that came together because they shared the same experience and wanted to do more than "just" individually donate money: "Even if it is just a little help for the victims from the distant Belgium." (Beru tsū 2011b) The impression of a "Japanese-helping-Japanese" event is also supported by the "Thank You Letter" (*Orei to hōkoku*) placed on the website (Beru Tsū 2011a).

The Triple Disaster – especially the nuclear catastrophe – resulted in large-scale protests and an increasingly political civil society in Japan. The diasporic community in Brussels, however, limited itself to support activities, without participating in or organizing social and political protests. Several characteristics of the diasporic community contributed to the lack of politicization. The protest in Japan following the Fukushima catastrophe was carried out and sustained by students and *furītā* (freelance or unemployed people) up to the age of 30 and the age group above 50.[28] The majority of the short-term Japanese nationals in Brussels, however, are career oriented and between 30 and 50 years old. This group in Brussels is also well integrated into a social network based on corporate affiliations (Nihonjinkai) with close ties to governmental institutions. Finally, the need to reintegrate after their return also meant that members of the community in Brussels did not take an openly critical position concerning the political establishment.[29] Age structure, employment situation, social as well as corporate bonds that discourage political activities, and the need to reintegrate can explain the lack of political protest in Brussels.

3.2 Activating Networks: Japanese Long-Term Female Residents in Ghent

When focusing on the long-term residents in the city of Ghent, we discovered that the four interviewees do not consider themselves as part of a general Japanese diaspora. They belong to different informal "Japanese" sub-groups, formed

---

[28] For the protests in Japan following the nuclear catastrophe, see Gengenbach and Trunk (2012).

[29] One interviewed short-term resident from Brussels commented on the decreasing application of Japanese teachers wanting to go abroad by giving the following reasons: low income, missing the development in Japan and easier life in Japan; Interview 15, October 24, 2012.

by the parameters of children (age and school), work place and social activities. These sub-groups form loose networks that are held together by the identity marker "homeland" and through events such as a New Year's dinner once a year, occasional "ladies nights", and activities with children.

All long-term interviewees in Ghent participated in several fundraising and support activities, usually events organized by Japanese friends or acquaintances. The members of the overlapping networks informed other members to spread the information on fundraising activities to such an extent that one interviewee from Ghent remarked: "It was impossible to go to all support events. There were just too many."[30] The activities organized were directed towards an "outside", non-Japanese audience – mainly Belgians who are in one way or another connected to Japan economically or culturally – for example, students of Japanese studies, martial arts clubs, people who went to a Japanese concert or movie. The activities of these groups show a high degree of cultural knowledge of the host country because they are using events and networks in daily Belgian social life. The support and fundraising events organized by the permanent Japanese citizens were not initiated by a given and well-defined group but by individuals functioning as *hubs*. These Japanese individuals used their contacts to activate and rely on informal networks that are not limited to Japanese nationals.

The support events as reaction to the shared experience of shock, on the one hand, strengthened the networking system of the Japanese citizens and brought the participants closer together (although just for a short period of time); on the other hand, pre-existing trenches were reaffirmed: "The support activities did not really change the community of Japanese in Ghent, but people who did not want to do anything gave a weird feeling and the friendship has cooled down. We didn't want to press. We are disappointed."[31]

In the following section, we will exemplify these findings by focusing on the activities and support events initiated by two female permanent residents in Ghent. The flea market sales action at the church of St. Jacob in Ghent is a traditional, semi-professional flea market with regular stands and is held every

---

[30] Interview 11, November 7, 2012.
[31] Interview 12, December 10, 2012.

week from Friday to Sunday; it is an excellent example of how the support actions of the permanent community were embedded into Belgian community life. A long-term, single, Japanese businesswoman in Ghent organized the charity sales action at this flea market, which was supported by other long-term female residents and their Belgian husbands. Selling at long-established, local markets requires a permit, which cannot be obtained easily. In the case of the support action, however, the organizer was able to use the reserved space of a friend and just needed a permit for this particular event. The products sold were donated by friends and acquaintances, contacted via e-mail, telephone, Twitter and Facebook, and included typical products that could be expected to be on sale at flea markets as well as products of Japanese origin. It was clearly marked that the products were sold for charity reasons and a charity box with a short explanatory text in Dutch and a *hinomaru* (flag of Japan) was placed on the sales stand as well. Because the regulations prohibit placing charity boxes on sales stands, a special permit was also obtained from the city hall. Using a traditional Belgian flea market for a support action requires cultural, structural and administrative understanding as well as an engagement with the city and its people. This kind of specific cultural knowledge, which also shows a certain degree of embeddedness into the host community and its social network, cannot be obtained by short-term residents. When arguing from the point of network analysis, we can state that the organizers were able to find a common code of cultural communication, a code that could be understood by the members of the home community.[32] The flea market itself was "broadcast" live on a video on the Ricobel blog, which is linked to the company of the organizer and meant for Japanese with an interest in Belgian culture.

Concerning the flea market in Ghent, the way the money was donated also shows a high degree of social network within the host community. Most charity actions donated the money directly to the Red Cross in Japan or Belgium. In the case of the flea market sale in Ghent, the organizers bought toys and sweets for children and sent these to the city of Kanazawa, from where they were finally forwarded to the areas hit by the triple disaster. Choosing Kanazawa seems odd

---

[32] For network analysis, see especially Blumer (1992).

at first because Kanazawa is about 420 kilometers away from Fukushima prefecture. However, the organizer was aware of the fact that Ghent and Kanazawa are
sister cities, and through personal contacts in Ghent's city hall, she was able to use
her social network for this action. In this sense, the action also strengthened the
Ghent-Kanazawa connection. Besides speaking of a communality of practice, it
is also important to mention the communication of practice, for example, the
dissemination and publication of support activities, fundraising and donations.
The gift-giving was also mentioned in two newspaper articles in the *Hokkoku
Shimbun*, a regional tabloid from Kanazawa. In the article, the toys and sweets
received were clearly marked as gifts from an individual Japanese female living
in Ghent, Belgium. The two headlines read: "From Belgium to the children of
the stricken area. Mrs. Ō from Ghent entrusts the city with toys and sweets"
and the second header refers to the gifts as "Belgian toys to the stricken areas.
Mrs. Ō from Ghent City. Entrusting sister city Kanazawa." The latter article also
shows two photos of the gifts being placed on a conference table and arranged
by a Caucasian woman (in one photo also watched by a Japanese woman) who,
although not identified in the article, is the Belgian Coordinator of International
Affairs at Kanazawa City Hall.

Support activities obviously have the intrinsic meaning of offering material
and moral relief to the disaster-stricken regions. However, the activities also
serve the secondary purpose of creating communality and belonging; therefore,
it is crucial to communicate the realized activities and the gift-giving of those
"inside" the community to those on the "outside".

An excellent example for the importance of communication is the produc-
tion of three videos by the above-mentioned businesswoman with messages of
support and sympathy for the victims in Japan.Thus, support activities were not
limited to providing financial and material support; they also extended to activities such as offering messages of solidarity that were meant to give moral
support. These messages were considered to be equally important as financial
or material support, which Ruth Martin also shows in her analysis of support
activities in London for the triple disaster. The people who appeared in the two
videos, "A support message from Belgium (1)/ (2) Be strong, Japan! (*Berugī kara*

東日本大震災

ベルギー玩具 被災地へ

ゲント市の大迫さん
姉妹都市金沢に寄託

東日本大震災で被災した子どもたちに送ろうと、ベルギー・ゲント市在住の雑貨店経営大迫徳子さん（＝千葉県出身＝）が、同市の姉妹都市・金沢市に玩具や菓子などの支援物資を寄せた。

今月三日に現地で開かれたのみの市で日用雑貨を並べ、その売上金約六万円でそろえた。コマ四十七個、塗り絵二十、ベルギーのチョコレート八十個、パズルセットなど計約二百点。

金沢市に姉妹都市から、被災地に向けた物資が送られてきたのは初めて。被災地への直接送付が困難だったことから、金沢市へ寄託。被災地や金沢市に避難してきた子どもたちに送る予定。

大迫さんは「普段見ないもの、目新しいもので、明るい気持ちになってほしい」とメッセージを寄せている。

山野之義市長の短文投稿サイト「ツイッター」に直接申し出た。

（村上一樹）

被災地の子どもたちに向けてベルギーから寄せられた菓子や玩具＝金沢市役所で

H23. 4.23　北國新聞
(2011)

ōen messēji [1]/ [2] Ganbare Nippon!)", were Ghent citizens, one city official, local traders or teachers, and students from Ghent University, all somehow connected to Japan or to the Japanese living in Ghent or visiting the city. Among them was the chocolatier on the Korenlei (a famous tourist spot), who counts many Japanese tourists among his clientele. The Let's Speak Japanese Club (*Nihongo hanasō kai*) from Antwerp also added a message of encouragement. The third video was a personal message from the mayor of Ghent, Daniel Termont. The oral English and Japanese messages were subtitled in Japanese and placed on YouTube as well as on the aforementioned Ricobel blog. The comments posted on both sites as well as numerous letters sent to Ghent University show that Japanese in the homeland indeed seemed to notice and appreciate the messages sent. Japanese

national TV station NHK also chose certain scenes for a sequel of support messages from all over the world that was shown regularly in commercial breaks the
months following the triple disaster.

The videos of support were born from the personal experience of shock, which
resulted in the urge to support the Japanese in the stricken areas.[33] However, the
videos can also be linked to the corporate world because the Ricobel blog, where
the videos were posted, is run by the Ricobel company, which is also mentioned
at the end of the video messages.

In the second example, another individual female employee in Ghent utilized
the work environment for organizing support activities. She functioned as hub in
at least two fundraising events. Inspired by the example of a Japanese colleague's
sister who folded and sold *tsuru* (origami cranes) in Japan, she and the colleague
participated in the Orizuru_PrayforJapan support activity and folded origami
cranes together with volunteering colleagues. On the Orizuru_PrayforJapan website (2011), the company receives special thanks from the organizers. The organizers visited other offices in their building and sold the cranes for the victims of
the disaster for the price of € 1. This activity was extended to the private context
by organizing tsuru activities at the schools of the interviewees' children and also
at the judo club of one of the children. The paper crane is a basic origami figure
that can easily be folded and is thus very well suited for a group activity involving
children. In one case, after obtaining permission from the school director, the
two Japanese women – dressed in kimonos – visited all classes in this school over
a period of two days and folded cranes together with the attending school children. The children then went home to sell the cranes to their parents, friends and
neighbors. In a second activity, the annual Japanese film festival Japan Square, of
which one interviewee is board member, served as platform for a tsuru sale action.[34] In both cases the organizers donated the money to the Red Cross Flanders
and also received permission to use the Red Cross logo (Orizuru_PrayForJapan
2011; Shinema tudei 2011).

---

[33] Interview 11, November 7, 2012.
[34] Interview 12, December 10, 2012. Interview 13, December 10 2012. On March 11, 2012 the film
festival also showed the film *A Sketch of Mujo*, a documentary on the Tōhoku disaster.

In contrast to the above-mentioned flea market activity, the tsuru activity was based on and made use of stereotypical and easily recognizable elements of Japanese culture for their fundraising activity, thereby making the activity readily identifiable as an activity related to Japan.[35] The crane is a bird of cultural significance in Japan, expressing hope, long life, recovery from illness and injury, peace, and now increasingly stands for solidarity. Through the tragic story of the Hiroshima atom bomb victim Sasaki Sadako, the crane is well known also in Belgium as a symbol of nuclear disarmament. The link with the nuclear catastrophe of Fukushima made the crane a symbol that could also be interpreted as protest against nuclear energy. The Japanese embassy also received a number of origami cranes (single or in a string of 1000 *senbazuru*), which were displayed in a special exhibition at the Cultural and Information Centre of the embassy as token of gratitude following the triple disaster.[36]

4 Practicing Communality

The support activities by Japanese nationals living in Belgium – long-term or short-term – were marked by the commitment to their home country. When Japanese artists organised exhibitions, classical concerts and other charity events under the name "ACT FOR JAPAN", they stated that the group's purpose is to provide relief to the "homeland":

> We are trying to unite our efforts and act to support our home country. From Belgium, where we are blessed with a rich cultural environment and opportunity to share our art, we pray for Japan's rebirth and declare the message that no matter how hard it may be, Japan will rise up and shine again, just like the sun always rises in the morning. In this time of darkness, Japan needs your

---

[35] This activity was referred to as "Japan Square-Tsuru Ladies" in the *Trade Flows & Cultural News*, the official magazine of the Belgium-Japan Association, Chamber of Commerce, and has a circulation of 2,500 print copies.

[36] Also other Japanese and Belgians who organized fundraising events and moral support activities also used the crane as trademark that could easily be recognized, for example, The *Dinner of the Rising Sun* in the Palace of the Colonies (Tervuren, Brussels) on May 26, 2011. For the price of € 200, which was donated to the Red Cross, guests were served a four-course dinner by Japanese and Belgian top chefs. The event was organized by a non-profit organization, whose representative is the sushi-chef of the Antwerp-based restaurant Ko'uzi, Kawada Junko, a long-term resident. The names of the chefs as well as the names of their restaurants were mentioned on the advertisement.

support. Until the day we overcome the difficulties and smile again, we will join our hands and act for Japan. (Act for Japan 2011)[37]

Long-term residents often tend to display a romantic, idealized, as well as stereotypical view of their home country, which becomes – borrowing a term Marilyn Ivy (1995) used – a kind of "internal exotic". For Singapore, Eyal Ben-Ari (2003: 128) states: "the Japanese expatriates discover their 'Japanese-ness'". The websites and flyers of support events display elements of a stereotypical and romantic (or even better, nostalgic) construction of Japan and Japaneseness. This practice of preserving and even exhibiting a distinctive identity vis-à-vis the host society can be linked to one of the core elements of a diaspora, which according to Brubaker is "boundary-maintenance" (Brubaker 2005). The triple disaster consequently increased a sense of being Japanese and the awareness of Japanese cultural identity, as one short-term resident from Brussels remarked:

> We became aware again about the bad and good sides of being Japanese. There are good things like immediate organization of reconstruction, but the chaos also let the dark sides of the people come to light, for example, profiting from the confusion, taking away or stealing things from others, not taking into consideration the situation at hand. We are used to thinking that these things do not exist in our society, but they do. The will to survive, to protect oneself, and to rescue was overwhelming. I think that the Japanese as a nation did not take enough responsibility for what has happened.[38]

The way the community tried to recover and reconstruct resulted in a feeling of togetherness and also in a promise to the future, based on the construction of ties between people.

> Even before [the triple disaster] I had met these people [who were organizing the flea market], but I also met other Japanese people through them, and by organizing this flea market together, it felt as though the power of solidarity (*danketsuryoku ga umareta*) as Japanese emerged.[39]

---

[37] The website also displays a message to the victims of the triple disaster in the Tōhoku region. It is striking that the term *furusato* is used when referring to Japan: "What had happened in our home(land)." The term "furusato", which Christine Yano (2002: 17) refers to as "the focal point of nostalgia and memory" refers, on the one hand, to one's (rural) place of birth and to the "idea of originary, emotive space and homeland", on the other hand (Robertson 1991: 5, 14).
[38] Interview 8, May 24, 2014.
[39] This quote was made by a short-term resident in Brussels and is taken from an online questionnaire conducted by BA and MA students at Ghent University in 2011.

This quote is especially striking because the phrase *danketsuryoku ga umareta*, which is translated here with the "power of solidarity emerged", can also mean the "ability to join activities emerged". This, first of all, refers to something that was felt to be absent thus far, namely a feeling of solidarity as a Japanese, a feeling that Japan lacked a civil society,[40] or as one interviewee put it: "Before I used to think that Japanese are egoistic, but this is changing now. They now know that we have to help each other. [...] They are more concerned. Their heart's eye has opened."[41]

Yet, where in Japan the term *kizuna* (binding) was used to describe the feeling of standing together after the catastrophe, none of the participants used this word in describing the feeling of solidarity. Even more, the above quote suggests a complex field of inclusion and exclusion because the speaker creates an emotional distance with the Japanese back "home" by using the grammatical markers of "I", "we" and "they".

5 Conclusion

We have argued that a group creates or re-confirms itself (from within) through practice. Despite the differences between both long-term and short-term diasporic communities, the triple disaster and its aftermath served as a catalyst for increased communication on a personal level within the community. As such, the (closed) community of short-term Japanese nationals strengthened and reinforced their ties within the Japanese community, while the permanent Japanese nationals in Belgium strengthened their contacts with other Japanese, and also reached out to Belgian nationals for participation in charity events. However, the strengthening of togetherness was – because the practice of support

---

[40] For the definition of the term "civil society" in Japan, see especially Ducke (2007). In Japan, the redefinition of the cultural and social system and the rediscovery of the civil society is coined in the term *kizuna* (bonds; connection), which was chosen to be the word of the year in 2011. For the controversial discussion of the term *kizuna* in Japan, see Tagsold (2012). Already in December 2011, a new political party named New Kizuna (Shintō Kizuna) was founded; it exploited the term *kizuna* for political ends. The party dissolved before the general election in December 2012. The main political aims were economic oriented and were directed against the proposal to raise the consumption tax as well as against the Trans-Pacific Partnership Agreement. See Kizuna Party.

[41] Interview 3, June 8, 2012.

was limited in time – only a temporary manifestation, following the first months after the triple disaster. The support events also served as a means to further define network boundaries because not all Japanese nationals were equally willing to participate in volunteer activities. The organization of support actions thus strengthened already-existing divisions and provided an acceptable reason for further alienation. Support activities not only strengthened the awareness of "Japaneseness" and of being Japanese, but also added – at least temporarily – the notion of a "civil society" as a new element of Japaneseness: the ability to stand together and support each other in times of catastrophe not only individually or locally but also nationally. In Japan, this feeling of solidarity also meant a politicization of civil society, whereas the support activities of the diasporic community in Belgium, in essence, remained apolitical and supportive of the existing political system.

Literature

ACT FOR JAPAN. 2011. http://www.actforjapan.be/"page_id=20. (accessed January 15, 2013)

ADSEI – ALGEMENE DIRECTIE STATISTIEK EN ECONOMISCHE INFORMATIE. 2010. *Bevolking per nationaliteit, geslacht, leeftijdsgroepen op 1/1/2010.* http://statbel.fgov.be/nl/modules/publications/statistiques/bevolking/population_natio_sexe_groupe_classe_d_ges_au_1er_janvier_2010.jsp. (accessed 13 January, 2013)

ADSEI – ALGEMENE DIRECTIE STATISTIEK EN ECONOMISCHE INFORMATIE. 2011. http://statbel.fgov.be/nl/modules/publications/statistiques/bevolking/downloads/bevolking_per_nationaliteit_per_gemeente_01-01-2011.jsp. (accessed January 13, 2013)

BELGIUM-JAPAN ASSOCIATION, CHAMBER OF COMMERCE. 2011. *Trade Flows & Cultural News.* 92.

BEFU, Harumi, "The Global Context of Japan Outside Japan." In: Befu, Harumi; Guichard-Anguis, Sylvie (eds.): *Globalizing Japan.* London, New York: Routledge: pp. 3–22.

BEN-ARI, Eyal. 2003. "The Japanese in Singapore: The Dynamics of an Expatriate Community." In: Goodman, Roger; Peach, Ceri; Takenaka, Ayumi; White, Paul (eds.): *Global Japan: The Experience of Japan's New Immigrant and Overseas Communities.* London, New York: Routledge, pp. 116–130.

BERU TSŪ. 2011a. *Tōhoku Kantō daishinsai ni tsuite.* http://www.bel2.jp/home/earthquake.html. (accessed January 15, 2013)

BERU TSŪ. 2011b. *Tōhoku chihō taiheiyō oki jishin chariti bazā.* http://www.bel2.jp/PNG/home/bazaar/bazaar%20flyer%20JP.pdf. (accessed January 15, 2013)

BESNIER, Niko; ROBBINS, Joel. 2009. "Equality, Inequality, and Exchange." In: *Journal of the Finnish Anthropological Society* 34 (4): pp. 71–80.

BLUMER, Herbert. 1992. "Der methodologische Standort des symbolischen Interaktionismus." In: Burkart, Roland; Hömberg, Walter (eds.): *Kommunikationstheorien: Ein Textbuch zur Einführung.* Wien: Wilhelm Braumüller, pp. 23–39.

BRUBAKER, Rogers. 2005. "The 'Diaspora' Diaspora." *Ethnic and Racial Studies* 28 (1): 1–19.

CENTRUM VOOR GELIJKHEID VAN KANSEN EN VOOR RACISMEBESTRIJDING. 2011. *Jaarverslag Migratie 2010.* Brussels.

CENTRUM VOOR GELIJKHEID VAN KANSEN EN VOOR RACISMEBESTRIJDING. 2012. *Jaarverslag Migratie 2011.* Brussels.

CWIERTKA Katarzyna. 2002. "Eating the Homeland: Japanese Expatriates in the Netherlands." In: Cwiertka, Katarzyna; Walraven, Boudewijn (eds.): *Asian Food. The Global and the Local.* London, New York: Routledge, pp. 133–152.

DUCKE, Isa. 2007. *Civil Society and the Internet in Japan.* London, New York: Routledge.

EUROSTAT. 2012. http://epp.eurostat.ec.europa.eu/portal/page/portal/population/data/database. (accessed January 13, 2013)

GENGENBACH Katrin; TRUNK, Maria. 2012. "Vor und nach >Fukushima<: Dynamiken sozialer Protestbewegungen in Japan seit der Jahrtausendwende." In: Chiavacci, David; Wieczorek, Iris (eds.): *Japan: Politik, Wirtschaft und Gesellschaft.* Berlin: VSJF, pp. 261–282.

GENT CITY. 2012. *Demografische gegevens 2011.* https://stad.gent/sites/default/files/page/docume nts/demo2011.pdf. (accessed January 15, 2013)

GLEBE, Günther. 2003. "Segregation and the Ethnoscape: The Japanese Business Community in Düsseldorf." In *Global Japan: The Experience of Japan's New Immigrant and Overseas Communities* Goodmann, Roger; Peach, Ceri; Takenaka, Ayumi; White, Paul (eds.) London, New York: Routledge, pp. 98–115.

HOFFMANN, Susanna M.; OLIVER-SMITH, Anthony. 2002. "Why Anthropologists Should Study Disaster." In: Hoffman, Susanna M.; Oliver-Smith, Anthony (eds.): *Catastrophe & Culture. The Anthropology of Disaster.* Santa Fe: School of American Research Press, pp. 3–22.

IVY, Marilyn. 1995. *Discourses of the Vanishing: Modernity Phantasm Japan.* Chicago, London: U of Chicago P.

JAPANESE SCHOOL OF BRUSSELS. /www.japanese-school-brussels.be/JSBHPenglish3.pdf. (accessed January 13, 2013)

KIZUNA PARTY. http://www.kizuna-party.jp/. (accessed January 15, 2013)

MISCNETTO. http://www.misc-net.com/. (accessed January 15, 2013)

ORIZURU PRAYFORJAPAN. 2011. http://orizuruprayforjapan.blogspot.be/. (accessed January 15, 2013)

PETITS-POIS. http://www.petits-pois.be/. (accessed January 15, 2013)

ROBERTSON, Jennifer. 1991. *Native and Newcomer: Making and Remaking a Japanese City*. Berkeley: U of California P.

SHINEMA TUDEI. 2011. *Berugī de "shinboru" "yattaman" nado wadaisaku o shūchū jōei no nihon eigasai kaisai"*. http://www.cinematoday.jp/page/N0031568. (accessed January 15, 2013)

TAGSOLD, Christian. 2011. "Establishing the Ideal Foreigner: Representations of the Japanese Community in Düsseldorf Germany." In: *Encounters* 3 (1): 143–168.

TAGSOLD, Christian. 2012. "Das Schriftzeichen des Jahres 2011 als Antwort auf das gefühlte Auseinanderbrechen der Gesellschaft." In: Chiavacci, David; Wieczorek, Iris (eds.): *Japan: Politik, Wirtschaft und Gesellschaft*. Berlin: VSJF, pp. 309–328.

THE MISSION OF JAPAN TO THE EUROPEAN UNION. 2011. *Countermeasures for the Great East Japan Earthquake: Briefing for European Companies and Business Organisations*. http://www.eu.emb-japan.go.jp/Countermeasures%20for%20the%20Great%20East%20Japan%20Earthquake.html. (accessed January 15, 2013)

TURNER, Edith. 2012. *Communitas: The Anthropology of Collective Joy*. New York: Palgrave Macmillan.

WHITE, Merry E. 1992. *The Japanese Overseas: Can They Go Home Again?* Princeton, NJ: Princeton UP.

WHITE, Paul. 2003. "The Japanese in London. From Transience to Settlement." In: Roger Goodman; Ceri Peach; Ayumi Takenaka; Paul White (eds.): *Global Japan: The Experience of Japan's New Immigrant and Overseas Communities*. London, New York: Routledge: pp. 79–97.

YANO, Christine R. 2002. *Tears of Longing. Nostalgia and the Nation in Japanese Popular Songs*. Cambridge, London: Harvard UP, 2002).

3/11 and the Japanese in London

Ruth Martin

On the morning of March 11 2011 devastating news from Japan reached London. What became known as the Great East Japan Earthquake had struck Tohoku at 2.46 pm – 6.46 am in London. The news had special significance for over 63,000 Japanese residents in the UK, as well as the community of "friends of Japan", which has built up as a result of diplomatic, business and personal links, especially over the past sixty years.

This chapter begins by explaining the presence of Japanese in the UK and the different groups that make up the Japanese community. It continues by outlining three phases following the disaster: The first was characterized by a state of shock and chaos during which time there seemed to be a coming together of the groups that have previously tended to be separate and hierarchically distinctive. A spirit of *communitas*, common following disaster, was evident (Turner 2012). The second was a compulsion to do something. At first to donate financially, or to organize fund raising activities which brought sections of the community together. A third was marked by a change in attitude towards donating and an apparent waning of the spirit of *communitas* as members of the community concentrated on the professions that had originally brought them to the UK. The chapter offers explanations and suggests lessons that may have been learnt. It points out the significant role played by Japanese who may previously have tended to operate on the periphery of the diplomatic and corporate world, including permanent residents. This is particularly relevant in terms of diaspora studies, along with the role of the UK as a welcoming host country.

1 Introduction: A Brief History and Explanation of Japanese Presence in the UK

The notion of global diaspora has been used in the context of migration over the past fifty to sixty years. It is particularly influenced by the global economy, characterised by the movement of transient professionals and their families and supported by those who make up the network of infrastructure that grows in order to support them. Anglo-Japanese relations are in fact long standing. 2013 saw events in the UK marking the 400th anniversary of the opening of trade, scientific, cultural and diplomatic ties between Britain and Japan. Relations had strengthened following the Treaty of Amity and Commerce (1858) and shortly after this the existing legations in each country were elevated to the status of Embassy. Britain supported the Japanese during the Russo Japanese War (1904-5) and in order to support the war efforts it was during this time that the first Japanese bank , the Yokohama Specie Bank, opened in London, subsequently playing a vital role in the development of the Japanese economic ties with London.

However, after the First World War, New York rather than London became the global financial centre, and in the 1920s and 30s economic friction developed between Britain and Japan. The setting up of a puppet state in Manchuria by the latter along with aggression in China led to growing British unease, with the two nations becoming enemies in 1941 during the Second World War. After the peace treaty of 1952, relations between the two were described simply as cold and suspicious' (Cortazzi 2001: xviii), and were hampered by memories of the British defeat in Singapore and the issue of prisoners of war.

After World War II Japanese banks returned to London, initially to help finance trade, and then from the 1960s to finance the expansion of Japanese manufacturing in the United Kingdom (ibid). This heralded a growth of the Japanese community in the UK when records of the number of Japanese residents registered with the Japanese Embassy in London were first published. Japanese investment in the UK grew in the 1970s, but it was during the 1980s that Japanese presence in the city of London grew to its height. This was due to a rapidly expanding Japanese economy, reinforced by strong credit ratings, large cash reserves generated by the high rate of Japanese consumer saving, and rising property prices (Newhall 1996; Sakai 2000).

By 1990, more than fifty Japanese financial institutions had established representative offices in London (Sakai 2000) and Japanese manufacturing increased accordingly. In 1987 there were 50 established manufacturing projects in the UK, by 1989 there were over 95 (Sasao 1990). In the 1980s all of the major Japanese consumer electrical companies had at least one plant in Britain, often sited in a regional development area, or a so-called New Town, such as Telford and Milton Keynes, where they received incentives from local British authorities. As of May 2011, shortly after the earthquake and tsunami, the UK was home to 120 European headquarters including Nissan, Honda and Toyota which manufacture 50 % of vehicles in the UK – more than 130,000 jobs are provided by Japanese companies. Japan is currently the number two investor in the UK, and with investments worth £ 22 billion, it represents the most important destination in the EU for Japan (Embassy of Japan 2011).

Accordingly, the population of Japanese living in the UK has steadily increased. In 1960, when records were first published by the Ministry of Foreign Affairs, 792 Japanese were registered with the Embassy of Japan in London as being residents in the UK. In 1970 the number had reached 2,806. By 1980 that number was 10,943; by 1990 44,351; by 2000 53,114 and in 2011 the number had risen to 63,011 (ibid). The UK is currently home to the largest number of Japanese in Europe and has the fourth largest population of Japanese overseas behind the United States, China and Australia. Registering at the embassy is not compulsory, so these figures may not represent the true number of Japanese living in the UK. The majority (58.2 percent) live in and around London where the head offices of many of the financial institutions and corporations are based.

The Japanese Ministry of Foreign Affairs divides this population according to purpose of stay: private company staff, journalists/press, self employed, and those working for government (these groups represent a largely transient population). In 2011 private company staff accounted for 28 % of the total figure, and adding the total of press, self employed, those working for government and others, the total percentage of this transient professional population for 2011 was 43 %. In addition, students/teachers/researchers represent a further 33 % of the total, with permanent residents representing 24 % (Embassy of Japan 2011; See Table 2).

Reactions to the disaster highlight four main players amongst this Japanese population that are the focus of this chapter: representatives of the government from the Japanese Embassy of Japan, the Japanese Corporate world, students, and permanent residents. The latter may be sub-divided into Japanese "entrepreneurs" who have broken away from the Japanese corporate world to set up their own independent businesses in the UK; those Japanese who are married to non-Japanese and now residing in the UK, and those who are based in the UK for the purpose of a particular talent , including musicians, dancers and artists. The latter represents an interesting development in Japanese transnational migration, meriting further attention in diaspora studies.

In comparison to some diaspora communities elsewhere, it is important to note that a defining feature of those making up the population of Japanese is their educated middle class and, although in the 1980s there was a feeling that those who had spent time overseas may not be quite Japanese' on their return, this group has come to represent a kind of elite in Japan. Studies of bankers have suggested that for many male managers, overseas transfer is an indication of future career prospects and the launch pad for future top managers (Hamada 1992; Morgan, Kelly, Sharpe and Whitely 2003). In the case of one particular bank for example, it was estimated that between twenty and thirty percent of the managing directors had international experience (Morgan, Kelly, Sharpe and Whitely 2003). This group of senior managers has contributed to a group of professionals in London which has often had more than one posting representing promotion on each occasion. These professionals have often developed close ties with the country – personal friends they have made through the Anglo-Japanese friendship groups mentioned below and within the business community. They also represent a privileged group of internationally-minded Japanese who come to move across the national boundaries between Japan and the UK with ease, including when their children remain in the UK to study, work or live upon marriage to non-Japanese after their own repatriation. As such, they seem to have a foot in both countries (Martin 2007) and the personal links they form during multiple postings is significant as this chapter will show.

Beneath the senior management level are those junior staff, and by definition their younger family members, whose experience in London may be very different from that of their *senpai* (seniors). Their overseas living allowances are less generous and, with the exception of mothers with young children at local non-

Japanese schools, they are more likely to be existing outside the Anglo-Japanese networks described below. They may live in areas in which there is a high density of other Japanese, with an infrastructure of food shops, travel and estate agents, insurance companies and a Japanese school nearby. They may have less chance to return to Japan during their transfer period than senior management and their point of reference to the homeland may therefore be different from their senior management colleagues.

The main groups within the Japanese community are quite distinctive and usually separate. The business community is particularly hierarchical and within it certain leading Japanese companies and their senior management play a pivotal role. These include Mitsubishi and Mitsui, reflecting two of the original "Big Four" corporate groups mentioned above. The heads of such companies in London, being head not only of the London branch but also in Europe, are at the top of the social hierarchy. This is evident in the network of Japanese business ties and subsequent social functions that ensue. They seem second only to the Japanese Ambassador and Senior Ministers of the Embassy of Japan, though in fact the links between the two may, by necessity, be close and characterized by mutual respect.

Of permanent residents, the majority of those Japanese women who have married non-Japanese men have tended to remain on the periphery of the social scene that revolves around the somewhat elite business world. The exception to this would be the rare cases of Japanese married to British men who are senior in the business or diplomatic world, often with a relationship to Japan themselves. There are also a number of entrepreneurs' some of whom first came to London as transient professionals themselves who have their own business interest in the UK and who retain status and links with the Japanese corporate world.

Side by side, and partly overlapping with the Japanese 'business and diplomatic elite' in London are the members of a similarly elite Anglo-Japanese community that has developed since diplomatic ties became formalized, made up of leading Japanese corporate members and their wives and of British 'friends of Japan', typically British who have worked in Japan as expatriates in important British companies in Japan, such as Shell, or for the diplomatic service. There are also scholars who were sent to Japan after the war who retain strong ties and research interests and who have played an important role in Anglo-Japanese understanding over the years. Some of these ties are based on years of professional and personal contact with Japan, building upon the personal links and friendships made over time as a result. They are often members of The Japan Society or of *Nichi-Ei Otomodachi Kai* (an Anglo-Japanese Friendship Group for women founded in 1962).

Here I focus on the Japanese reactions to the disaster and on the reactions of the Anglo-Japanese community in London, as the financial and business centre, accounting for the densest population of Japanese not only in the UK, but also in Europe. The more elite business world and the related Anglo-Japanese friendship network discussed above are axiomatically centred here. However, following the disaster, fund raising activities took place all over the country and support came from all sections of British society: Japanese Embassy staff in London were faced with a huge work load, personally attending as many of the fund raising events as possible, sometimes several events in a day.

2   The First Phase: Shock and Emotion in the Immediate Aftermath

The first three months were characterized by shock. Japanese ballerina Miyako Yoshida, who was then based in London, summed up well the experience of many Japanese residents in the UK on hearing the news of the disaster: "I first found out about it on the BBC news. It's difficult to describe. I was shaking... I felt the blood drain from my face... I was panicking..... I immediately contacted Japan, and got in touch with my family and friends." (Yoshida 2011)

The Embassy of Japan in London became the focal point for the deluge enquiries from Japanese residents concerned about family members and friends in Japan, as well as about the safety of the Fukushima Daiichi Nuclear Power Plant. The Embassy seemed to many the obvious first point of call and its role as representative of the Japanese government was paramount in the immediate aftermath. These were additional to diplomatic roles that needed to be carried out, including managing the departure of the envoys that had been offered by the British Government. Only one day after the disaster, Britain was the first to send out a search and rescue team, and so began the job for Embassy staff of making arrangements for the team and their two dogs. The team consisted of 59 specialists, two rescue dogs and four medical staff, who spent four days searching the towns of Ofun and Kamaishi following the earthquake and tsunami. The specialists came from Lancashire, Lincolnshire. Greater Manchester, West Sussex, Kent, West midlands, Mid and West Wales, Hertfordshire and Cheshire. As well as the rescue team, the United Kingdom government offered 100 tons of drinking water to Ibaraki Prefecture and on April 2 provided nuclear related-equipment (radiation survey meters, protective masks etc.).

Another role carried out by the Embassy in its diplomatic capacity was to open a book of condolences. This allowed both Japanese citizens and members of the British public to express their solidarity with and sympathy for Japan on paper – a mainly symbolic gesture, but the act of expression felt important to those who participated and did not go unnoticed by Diplomatic staff. Between March 1 to 23 over 500 people visited the Embassy to sign , including the Prime Minister David Cameron, Foreign Secretary William Hague and the Duchess and Duke of Gloucester; the Duke has strong links with Japan, is patron of the Japan Society, and on this occasion signed on behalf of Her Majesty the Queen. In real terms, however, the number signing represents a tiny fraction of Japanese residents in the UK and did not represent an adequate or realistic gesture for many, who remained unsure of what to do, how to contribute or how to express their emotion and concern.

During this time, one London-based Japanese musician in particular was able to capture the spirit of fellow Japanese and provide a platform for them, not only to give financially in the collection buckets provided, but perhaps more importantly, to gather together in one place and show solidarity for Japan. His first appearance was only three days after the earthquake when he played at Mitsukoshi Department Store in London's Regent Street. Some 500, mainly Japanese, gathered to hear him and around £ 15,000 was collected. Another concert was held in the Fortnum and Mason store and he also busked in Kensington High Street and London St Pancras Station. In the early days, letting Japan know that Japanese overseas were expressing solidarity was as vital as any fund raising effort. The message seemed to be, 'We are all the same Japanese identity', or *onaji nihonjin dōshi.*

Such reactions appeared to demonstrate what Edith Turner refers to as the *communitas of disaster* (2012: 76) and how sharing a common experience can bring individuals onto an equal level (Turner 1969: 132). Japanese responses clearly demonstrated this levelling effect amongst the three groups which were normally clearly defined hierarchically. Superficially at least, the period appeared to show all sections of a community that is normally defined by strict hierarchy, marked by the social groups outlined above, to be united in a spirit of 'affect' or emotion' (see Besnier's Introduction).

A key factor that united Japanese residents during this time was being Japanese. A study of the wives of Japanese professionals in London showed how temporary residence in the UK leads to an appreciation of both the positive and negative aspects of the host country, which can in turn lead to a heightened sense of home (Martin 2007). Japanese mothers reported that they felt an increased burden to pass on a knowledge and awareness of their Japanese identity to their children in the same way that their own mothers had in Japan (ibid). As Befu states, 'sometimes expatriates are the most ardent patriots' (2001: 16) and perhaps no more so than at a time of disaster.

Another factor unifying the Japanese populations was concern about family members in Japan (expressed in the enquiries to the Embassy above). It is particularly pertinent for transient professionals for whom a temporary job transfer may have meant leaving children behind at school, university or work in Japan, and/or elderly parents. With concerns about radiation from Fukushima, and stories in London of water and power cuts along with the fear of aftershocks, the desire to call such children to join them to safety in London was strong, but mixed with feelings of guilt at the thought of doing so. In line with many expatriate British working in Tokyo at the time, children and family members themselves commonly felt that they could not leave Japan. There was also a corresponding feeling of guilt amongst some Japanese in London that they were safe while fellow Japanese were suffering. The issue of guilt suffered when 'spared' from natural disaster to the Holocaust is of course the focus of a whole body of literature (e. g. Lifton 1993).

A further unifying factor was the realization that an earthquake, which every Japanese citizen has prepared for since drills at kindergarten, had actually happened. And not only that it had occurred but that they were not there at the time and 'It could have been me.' Moreover, since the possibility of future earthquakes remain, It could still be me'.

In the aftermath, worldwide media focussed keenly on the resilience of the Japanese people, and for Japanese residents in London, there was a pride and almost comfort in this '*gambarou*' spirit which further united them.

3   The Second Phase: The Urge to Do Something

Out of the shock rose the urge to do something. Besnier reminds us that here is a strong link between affect and action, and *communitas* has been argued to represent activity rather than a state (Turner 2012: 221). The compulsion to act was expressed firstly in terms of making financial donations and of either organizing or attending the plethora of fund raising activities held all over the country. The mantra of "donations, donations, donations" along with the organization of charity events to provide such donations was dominant.

From the time the news broke, in addition to dealing with the mass of enquiries mentioned above, a further major task for the Embassy of Japan had been to respond to the financial donations from both Japanese residents and the British alike, perhaps outside the normal remit of its diplomatic role. Due to massive demand, the Embassy started accepting donations to be forwarded to the Japanese Red Cross. The Red Cross Fund eventually raised an estimated £ 14.2 million world-wide, which was essential in the immediate aftermath to provide urgent medical care and psychosocial support; to distribute blankets, clothing and other relief items; to support social welfare services; construct temporary hospitals; support children's education; improve the living conditions of people in evacuation centers or temporary housing, and to help people affected by the nuclear power plant disaster (BritishRedCross 2011). The Embassy also

69

offered a further service: to list, and thereby publicize, all events held by various groups and organizations in aid of the disaster relief.

Fund raising took the form of bazaars, auctions, concerts, street collections and more. Japanese mothers and young children as well as university students arranged events. Donations came from all sections of the community – from Japanese mothers and their children, school children, groups and organizations, Japanese companies and from corresponding initiatives from British citizens. Japanese Embassy staff in London were faced with a huge work load, personally attending as many of the fund raising events as possible, sometimes several events in a day in order to represent the support and appreciation of the Japanese government. The Ambassador himself hosted a bazaar at his official residence, which was supported by the Embassy and organized mainly by *Eikoku Fujin Kai* (The Japanese Women's Association in Great Britain). Attended by more than 450 invited guests, including representative of both the Japanese and British communities as well as those from the diplomatic corps in London, the amount of sales on the day was over £ 40,000. The total sum raised by donations was remarkably in six figures - over £ 250,000 - Japanese corporations and individuals having also donated generously.

The urge to do something resulted in members of the different groups coming into contact with those from other groups that they may not have otherwise had contact with. Embassy staff said that they met individuals outside their usual network that they would not otherwise have come in contact with. Through word of mouth and social networks including Facebook, other Japanese residents began to discover fellow Japanese living or working nearby, for example students discovered fellow Japanese at their university or college they never knew existed. They began to gather to discuss organizing bucket collections and other activities on campus. Annual events held at the Embassy since the disaster, such as the Emperor's birthday celebration, seemed to have hosted a greater number and wider range of guests in reflection of the number and range of individuals met through their contribution to the relief aid.

Several other factors accelerated the apparent breaking down of the formal social hierarchy within the Japanese community. First was the fact that the aftermath – especially the first three months – was a period of relative chaos, and the emergency of the situation meant that there was no time for the usual Japanese lengthy decision-making process and there was no comparable precedent to follow since the Kobe earthquake of 1995. Japanese residents quickly relied on new social networks, such as Facebook, both to gain information about the situation in Tohoku and subsequently to disseminate information on fund raising events in the UK. These social media are not respecters of any social hierarchy that allows business and diplomatic circles to have privileged information of, and access to, events or information.

For example, news of a performance organized by ballerina Miyako Yoshida at the Royal Opera House was spread by the internet, – on ballet blogs and social networking sites – and by word of mouth. It sold out in days without any other advertising means. The wife of the London and European Head of a Japanese company, who might normally have enjoyed seeing Miyako from the stalls of the Opera House thanks to *settai* (client entertaining) was to be seen queuing for returns – which she did not get. Furthermore, she did not seem to mind: she was there and the cause was more important than seeing the performance. Normally within the Japanese community, events would require planning and would go through appropriate networks of communication. Events such as the ballet gala were organized at short notice, bypassing the accepted social etiquette that is particularly defined amongst the diplomatic and business networks.

The relatively short notice at which a memorial service was organized at Westminster Abbey on Sunday, June 5th, just under three months after the disaster was another example of breaking down social hierarchy and of the spirit of *communitas*. Anyone was entitled to apply for tickets from the Westminster Abbey office and within the Abbey, seating seemed to be relatively free and removed from the normal constraints of etiquette that may have seen the Ambassador and heads of major companies seated accordingly and greeted with ceremony. The service itself reflected all sections of the Japanese community and friends of Japan, from the Ambassador to Japanese school children and, if anything, was orchestrated by certain permanent residents, who may normally be on the periphery of the hierarchical business and diplomatic world, but who had contacts with the church in the UK or with musical community that contributed to the service. Contributions to the service indicated a collaboration between all sections of the community as well as the strength of Anglo-Japanese links, such as an address by the British church minister who had worked as a minister in Japan, and a testimony by a Japanese Professor of Ophthalmology currently working at University College London. In addition to the Christian service and hymns, Buddhist monks offered prayers for the living and the dead, and traditional Japanese music was played at the end of the service. In an unprecedented move the Japanese flag was flown above the Abbey – somewhat symbolic in view of past war-related memories touched upon in the Introduction. The mood therefore seemed to be one of unity – between all sections of the Japanese community and between the UK and Japan.

4   The Third Phase: Changing Attitudes to Financial Giving and the Return to Daily Life in London

As time went on it was inevitable that enthusiasm for financial donation waned as residents went about the business of their daily lives, particularly in the case of transient professionals and their families who tend to feel strong obligation to their work and very purpose for being in the UK. This is especially the case of more middle of junior management who are particularly conscious of their obligations to the company they represent. There is a limit to how much an individual can financially donate and to how many events they can attend. After a while people naturally began to tire of the plethora of charity events, or felt obliged to attend simply because they were Japanese. As residents attended to their daily business, a return to the hierarchies of the corporate and diplomatic worlds that are defining features of the diaspora in question was inevitable.

People also wondered what had happened to the amounts of money raised. Students who had gathered to collect in buckets on University campuses could not see what had happened to the money. They could not be sure that their hard work had actually done good. With so many separate groups collecting, there was also concern whether money was getting to where it was meant to, or where it was needed most, and however misguided, some also expressed dissatisfaction with the Red Cross, suspicious that with such a large charitable organization, much would be spent on administration costs rather directly to the cause. Some worried that their contributions would go to other projects within the Red Cross umbrella rather than to Tohoku. All of this was contributed to as realization that perhaps something more than financial donations was needed to help victims, support local businesses and to rebuild the local economy for the long term.

In setting up its own fund, (The Tohoku Earthquake Relief Fund), The Japan Society Fund had been quick to realize the importance of liaison with local voluntary, community and non-profit organizations (NPOs). Other Japanese groups and individuals in the UK began to realize that more emphasis was now needed on this approach, rather than on just giving financially, and to realise that to contributions in the form of offers of help, or of *pro bono* donations in terms of goods or services, might be more useful. In addition it became clearer to all that action must be focussed on the needs of victims in the stricken areas, on what they needed to rebuild their lives for the immediate and long term, particularly in practical terms such as in the building up of local businesses and the local economy.

The EWWA (East West Art Award) art group for example, arranged for profits from its art exhibition to be spent on art materials rather than just sending the funds to Japan, with the aim of giving individuals in Tohoku vouchers to obtain these materials from art shops in the area. Such initiatives help not only those who receive the art materials, but the businesses who normally sell them. But caution was needed: sending bikes to Tohoku for example may help the recipients, but not so helpful for any local business in Tohoku who would normally have sold bikes.

The Tohoku Earthquake Relief Project (TERP) is a London-based group that has encouraged the above approach, seeking to act as a platform for introducing individuals or groups in the UK to projects they may best be suited to help with in Tohoku. Projects have included the Wallpaper Project, which was set up by London-based interior designer Noriko Sawayama who provided high-quality British wallpaper for the temporary housing built for victims. The wallpaper makes a difference to the quality of life of victims residing in such accommodation for much longer than was originally anticipated. Several similar projects have been set up through TERP: A group of Japanese mothers in the UK set up The Baby Muslin project which provide squares of muslin for mothers and babies or small children in the stricken area; East Loop aims to create business opportunities within the area by enabling them to sell their own handcrafts for which they then keep the profits; Seven Beach Aid supports the Seven Beach (Shichigahama) area in Miyagi Prefecture and has organized photographic exhibitions to show progress in the area; Smile Kids has been working specifically to support orphanages, and Sono x Nadeshiko was set up by a fashion designer who collaborates with women in the area to make fashion accessories from kimono material salvaged from the tsunami.

The approach shown in such projects enables or encourages people-to-people links between the benefactors and beneficiaries, and most importantly of all enables dialogue between the two to ensure that beneficiaries receive the help they truly need rather than what benefactors may presume. Such projects demonstrate *communitas* between the individuals involved, but require organization and commitment that can be less easy to maintain over time, especially in view of the nature of commitment required by transient staff to their posting, which may be another factor in the waning of the initial 'disaster spirit'.

Another lesson learnt in this is the importance of feedback. One of the reasons for public malaise in giving financially was lack of feedback about what had happened to their donations. The experience of TERP has shown how individuals who may be able to help fall into different camps: from those who have a ready made skill; business idea that can be linked to the area (such as in the case of the wallpaper project); individuals who were keen to help but don't know what to do or how, and those who say they are interested but too busy with their daily commitments. It seems not only important but crucial to tap into all of these and recognize that each may have a contribution on whatever scale. TERP therefore organizes *renraku kyogi kai* (information meetings) and 'volunteer pub' – informal occasions where interested individuals can meet and hear about ongoing projects and their progress. It also has a Facebook profile and sends out regular newsletters by email. The target is for 500 readers, accounting for 1/100 Japanese in UK. So far the newsletters have around 300 readers and the Facebook page 400 followers.

One point that TERP has discovered in trying to put people's skills to use is that it is more helpful if volunteers can come and say, "This is what we can do" rather than, "What can we do". Furthermore, while there will be individuals who have definite skills and connections in Japan they are able to put into action (such as the wallpaper project), there will be other individuals who, while they would like to help, don't know how to, or those whose time is limited to work or family commitments but would like to help in all ways if they can. Volunteering has been the subject of research since the Kobe (The Great Hanshin) Earthquake of 1995 or even before, but it may be that Japanese approaches toward charitable organizations could to be reassessed long with education about the notion of charity organizations, charitable giving and volunteering in Japan.

5 Youth and Social Media

The role of young people and the way they have gained, developed and employed skills in charitable organization is a significant factor to emerge from the disaster. Previously on the periphery of hierarchical social networks in the UK, students were not only part of the first phase of *communitas*, they were at times driving forces of the second phase.

It was young musicians who took initiative from Taro Hakase's performances not only to spread word of his appearances and to man collection buckets, but to follow his example and organize fund raising concerts of their own. Young musicians studying at London's music conservatoires were also invited to play at more elite fund raising events, once again demonstrating a bringing together of members of the different groups. Young Japanese were key in a number of proactive groups, such as Play For Japan, which brings together young professionals, students, young musicians and artists with the aim of using their skills to raise the profile of fund raising events and of sustaining awareness of the continuing challenges being faced in Japan during reconstruction. The organization received support from the Japanese Embassy of Japan as well as the Japan Foundation and the Japan Society, showing once more the value of cooperation between all three. Other groups include Young- Kai, set up by students studying in London, who held a candle-lit ceremony on the anniversary of the disaster and a photographic exhibition called One Second and Action for Japan which organizes both charity events and volunteer tours to Tohoku for British students.

Young people have been particularly good at employing social networks such as Facebook, as well as e-mail and the Internet and in a way which some extent superseded other forms of media such as TV and newspaper coverage. Reporting of reactions to the disaster in the UK by the more traditional Japanese media of television and the press has been aimed at emphasizing support for Japan from Japanese living in London as well as the host country and promoting the message, "You are not alone". This was seen in reporting of the concerts held by Takase in the aftermath and in reporting of the event at the Royal Opera House.

Disadvantages of television and press coverage included the inescapable fact that some projects are more attractive than others for their visual and symbolic appeal. The 'Sakura Front' project for example, received much media attention since it involved making poppies for remembrance, with half of the sale profits sent to Tohoku where they were used to plant *sakura* (cherry blossom) trees. Organizers of some charitable projects found that the presence of television cameras could bring tensions when individual victims or groups were singled out and there has been concern that where cameras go in and film, they tend to be focused on a "good story" and aftercare and other needs of victims are not a priority.

Reporting by the British media in the UK gradually decreased with time, apart from regular updates on the nuclear-related concerns at Fukushima. The first and second anniversaries of the disaster were opportunities for the media to bring the event back to the attention of the British public. The Japanese Ambassador used the occasion of the first anniversary to publish a full page message in the Independent on Sunday newspaper. This was headed '*Arigatou, Eikoku. Ganbare, Nihon*' (Thank you, UK. Fight on, Tohoku) – a play on words of the newspaper's own front page message two days after the disaster, '*Ganbare, Nihon. Ganbare, Tohoku*' (Don't give up, Japan. Don't give up, Tohoku). Both messages were also well reported in Japan, once again reflecting the use of more traditional media as an expression of the message of support .

Forms of social media such as Facebook and Twitter have obvious advantages, not least immediacy and the ability to reach a wide range of people on an equal level. Social media has allowed individuals to find out about the ongoing situation in Tohoku, as well as to be kept informed about fund raising events in the UK and to publicize them. The use of social networks has led not only to the inclusion of students in activities for which they may formerly have been on the periphery, but has meant that they have some control over such activities themselves. This has been a major contributory factor in breaking down social hierarchies and in enabling all sections of the Japanese community including young people to have a voice.

6   The Increasing Role of Permanent Residents

Reactions to the disaster highlight the extent to which Japanese permanent residents have gradually become established in the UK, perhaps corroborating the observation of Marienstras (1989: 125) that the reality of a diaspora is 'proved in time and tested by time'.

While the 1960s saw the foundation of Japanese groups aimed mainly at the transient professional community like the Nippon Club, JCCI (the Japanese Chamber of Commerce and Industry) and *Eikoku Fujin Kai* (Japanese Women's Association of Great Britain), an organization for permanent residents was not formed until some thirty year later in 1991 – the Japanese Residents Association, which changed its name to Japan Association in 1997, and was registered as a charity in 2011. The Association reflects the needs of the increasingly long established permanent population, especially as they reached retirement age, some having lived in the UK for up to 50 years as a result of finance and manufacturing, and who as already mentioned originally might have been transient professionals themselves. It focusses on mutual support, health and welfare of its members, for example by visiting those who are sick. Monthly support meetings attract up to 80 members. Members also maintain a Japanese cemetery, which was first established in 1936 but fell into disrepair following World War II. The Association arranges lecture meetings, a walking group and the promotion of Anglo-Japanese understanding and cultural activities.

It is significant that the Japan Association promotes volunteering, charity activities and cultural exchange for those living in the UK and that current membership is not exclusive to Japanese permanent residents of the UK but open to anyone who speaks Japanese. This expresses inclusivity in contrast to the exclusivity often conveyed by the more elite business transient world, and a breaking away from the mould of Japanese traditional corporate life and its etiquette, while at the same time maintaining the ability to operate within it. The Residents Association has also come to understand the value of cooperation with other organizations, both British and Japanese, such at the Japan Society, JCCI and the Nippon Club as well as with the Embassy of Japan from whom it has gained respect through its actions – for example in the organization of the *matsuri* that has been held for the past three years.

Key members of the Residents Association were particularly involved in this. Such entrepreneurs who are particularly well connected as a result of their own business experience in the UK can be highly regarded rather than looked upon with suspicion as may have been the case in the past, particularly the 1970s and 1980s, when they may have been thought of as "not quite Japanese". Like those female "local staff" working in Japanese Banks in the City of London described by Sakai (2000), permanent residents often become a resource for information and can be respected by the transient population for their knowledge of life in the UK and as such, often form a significant bridge between the two cultures. Permanent residents have come to play such a significant role precisely because they are not stuck within the confines of the Japanese corporate and diplomatic world but have experience and understanding of it: they maintain connections with its members and hold their respect but are able to draw upon both.

Thus while there may have been a view that permanent residents were on the periphery of the business and diplomatic social world, such residents have come to play an increasingly important role. While the transient population who help make up the diaspora come and go (and sometimes come back again) permanent residents add a permanency to the community entered into and have much to teach the wider diaspora community about promoting connections between all groups that make up the community. They also have the Anglo-Japanese connections that have proved vital in contributing to disaster relief.

7   Responses to the Disaster and Anglo-Japanese Relations

Closely related to the above, responses to the disaster in Japan have also illustrated the strength of Anglo-Japanese relations and have further contributed to them. Strong links between the home and the host nation can be vital in times of disaster. In terms of diaspora, it is relevant that the UK is a welcoming and tolerant host country, one of the characteristics that Cohen suggests as significant in allowing the possibility of a distinctive, creative and enriching life (Cohen 1997: 180).

Interest in Japan amongst the public was strengthened by two nationwide Japan Festivals in 1991 and 2002 and evident in the 450 or more events that took place throughout the entire country in 2008 to mark the 150th anniversary of formal diplomatic relations collectively referred to as 'JAPAN-UK 150'. Public interest shown in the Matsuri mentioned above surpassed the organizers' wildest dreams and support for Japan following the disaster of 2011 came from all corners of the UK and from all manner of people.

Since diplomatic relations were formalized, personal links that have developed, especially in the last fifty to sixty years, have also formed an influential community of friends of Japan that have played an important role in Anglo-Japanese relations. These are typified by the Japan Society and *Nichi-Ei Otomodachi Kai*, the Anglo-Japanese Friendship Group for the wives of senior Japanese business men and Diplomats who are matched with a British member to act as friend (*otomodachi*) during their time in London. The Group was founded in 1961 and over the years some Japanese members have rejoined on second postings to London, while British members develop strong ties with their Japanese "friends". Such links have been important in the aftermath of the disaster, for example in terms of the connections that the Japan Society was able to use in administering its Relief Fund.

Strong relations between members of the Royal and Imperial families and the similarities between the two island nations that make the UK a welcoming host country have been important in promoting Anglo-Japanese relations and an understanding of Japan in the UK. There has been a strong tradition of Japanese princes being educated at Oxford, including the current Crown Prince. The year following the disaster coincided with a British Royal event that further linked the two countries – celebrations marking sixty years since the coronation of Queen Elizabeth II, which was attended by the Emperor of Japan as one of only two living royal family members who had been present at the coronation itself. The Emperor and Empress arrived in London in May 2012 for a five-day state visit which was used as an opportunity to thank not only the British people but also members of the Japanese community, commenting not only on its contribution to fundraising, but significantly, on its role in building Anglo-Japanese relations over the years. In this the Emperor acknowledged an important function of diaspora communities, especially at times of disaster.

The importance of Anglo-Japanese ties was seen nowhere better than in the response to the disaster by The Japan Society, which was able to use contacts built up over time to work closely with the Sanburi Foundation in Sendai in order to best support local relief activity in the region. As of March 14 2012 the total raised was £677,000 from 1600 donors (Japan Society 2012). Although a British organization, the Society has a Japanese membership and is particularly generously supported by Japanese corporations in the UK, marking a strong and essential Japanese involvement. In addition, Japanese women were instrumental in translating the letters from school children in the school project set up by the Society following the disaster. Such collaboration between both individuals and organizations of host and home nation demonstrates the value of Anglo-Japanese links as a lesson for diaspora communities.

8   Conclusion

The first phase of *communitas* seen in the aftermath of the disaster was characterized by emotion and by a consequent levelling between members of the groups that are normally separate and differentiated. The second phase following the initial shock was associated with the desire to do something. Initially this was reflected in a determination to donate financially or to organize or support events that were held to raise funds. However, over time there became a wider understanding that, though such funds were essential in the aftermath, a different approach was needed for the long term recovery of the local economy and that the needs of victims themselves needed to be heard. Along with the return to the normality of daily life in London this contributed to a natural dissolution of the original enthusiasm as members of each group returned to the roles that define their very existence as part of the Japanese diaspora in the UK.

In this way, the aftermath of the disaster has perhaps shown that the existence of groups within the community of Japanese can have both merit and purpose. In the sense that social networks can have value (Putnam 1995), their organization may be said to have social capital. The diplomatic and practical function of the Embassy, for example, has been vital. In addition to the functions mentioned, the Ambassador used the occasions of ceremonies of remembrance hosted at the Embassy one and two years after the tragedy as an opportunity to raise the issue of the European Union's blanket restrictions on imports from Japan, to promote the produce of Tohoku and the safety of travel to the area. Conveying such messages was an important diplomatic role of the Embassy, and yet in terms of effective responses in the aftermath, diplomatic staff did not act alone.

Side by side with the diplomatic community, the business community played its own vital role, particularly in terms of being in a position to make large financial donations, and through their sponsorship of the Japan Society they supported the Japan Society Relief Fund with its focus on the work of NPOs. Similarly, the contribution of students and permanent residents is also clear: this chapter has shown how students and permanent residents normally on the periphery of social networks can not only be drawn into spirit of *communitas* following disaster, but they can also become a driving force of it. The use of social media not only facilitated their inclusion, but gave them a considerable amount of control over it.

Students and permanent residents can perhaps become driving forces precisely because they are not limited by restrictions found in other hierarchical social groups, and yet at the same time, the connections found within these groups can be essential in getting things done, thereby paradoxically reenforcing their existence. The musicians and performers who were instrumental in providing the platform for other Japanese to express solidarity were high profile themselves, already known and with ready-made connections in business or diplomatic circles. It was Takase's celebrity status that enabled him to arrange the series of performances at short notice at a time when fellow Japanese in London did not know how to express their emotion or know what to do. Paradoxically, rather than being broken down as the initial phase of apparent *communitas* might have suggested, the hierarchical system was perhaps as important as ever.

Responses to the disaster have shown how permanent residents have become more established in the UK, perhaps justifying the term diaspora, and how they too can play a vital role within it. Permanent residents maintain the very permanence of the community into which transient professional and their families enter and at times of disaster this permanence can have particular value. They are not merely focussed on their professional roles but can have a wider outlook and play important parts in bridging between the host and the home nation and its transient professionals. This has been seen in the support given by certain permanent residents – especially in the Residents Association – to the Japan Society and in arranging events like the *Matsuri*, which involves cooperation not just with the Japan Society but with organizations for the transient corporate world (such as JCCI and the Nippon Club).

The value of good relations with the host country built up over time is also demonstrated, especially on the personal level. In setting up its Relief Fund, the Japan Society was able to use the personal connections built up over the years. Such connections can be vital and can also reinforce the conventional hierarchical system.

Following the initial fervor to donate money and to organize or attend charity events the experience of organizations such as TERP seems to suggest a gap between attitudes towards charity in the UK and Japan. Whereas many British charitable organizations are well managed, even with potential political influence, there tends to be a different view to charitable giving in Japan. Volunteering at the local level is well established in Japan, such as in neighborhood organizations and PTAs (Parent Teacher Associations) at schools. However, the experience of students in becoming involved in organized charitable work in the UK as a result of the disaster may have provided lessons amongst a younger potentially influential generation. The skills learnt in the aftermath may be usefully employed in changing attitudes and in understanding of sustained charitable giving and of charitable administration that may help both affected areas long term as well as future disaster response planning.

Since diplomatic relations were formalized, personal links – especially at the corporate and diplomatic level including those amongst members of the Japan Society and *Nichi-Ei Otomodachi Kai* – have formed an influential community of friends of Japan that have played an important role in Anglo-Japanese relations. These, along with the similarities between the two nations that make the UK a welcoming host country, have been important in promoting Anglo-Japanese relations and an understanding of Japan in the UK. Responses shown in the aftermath of the disaster of 2011 demonstrate the strength of these relations, building on the 150 years of Anglo-Japanese diplomacy that, along with the nature of the global economy, facilitated the Japanese presence in the UK in the first place, and prove the durability of the Japanese diaspora in the UK, simultaneously demonstrating the value of diaspora communities in times of disaster.

Literature

BEFU, Harumi. 2001. "The Global Context of Japan Outside Japan". In: Befu, Harumi; Guichard-Anguis, Sylvie (eds.): *Globalizing Japan: Ethnography of the Japanese Presence in Asia, Europe and America.* London and New York: Routledge, pp. 3–22.

BRITISHREDCROSS. 2011. Japan Tsunami Appeal 2011. http://www.redcross.org.uk/What-we-do/Emergency-response/Past-emergency-appeals/Japan-Earthquake-Appeal-2011. (accessed October 17, 2012)

COHEN, Robin. 1997. *Global Diasporas: An Introduction.* London: UCL Press.

CORTTAZZI, Hugh (ed.). 2001. *Japan Experiences: Fifty Years, One Hundred Views.* Richmond, Surrey: Japan Library, Curzon Press Ltd.

EMBASSY OF JAPAN. 2011. Eikoku ni okeru zairyūhōjinsū. http://www.uk.emb-japan.go.jp/jp/ryoji/tokei.html. (accessed October 17, 2012)

HAMADA, Tomoko. 1992. "Under the Silk Banner: The Japanese Company and its Overseas Managers". In: Lebra, Takie Sugiyama (ed.): *Japanese Social Organizations*. Honolulu: U of Hawaii Press, pp. 135–165.

JAPAN SOCIETY. 2012. Japan Earthquake Relief Fund. http://japansociety.org/page/earthq uake. (accessed October 17, 2012)

LIFTON, Robert Jay. 1993. "From Hiroshima to Nazi doctors: The Evolution of Psychoformative Approaches to Understanding Traumatic Stress Syndromes". In: Wilson, John P.; Raphael, Beverly (eds.).: *International Handbook of Traumatic Stress Syndromes*. New York: Plenum Press, pp. 11–23.

MARIENSTRAS, Richard. 1989. "On the Notion Diaspora". In: Chaliand, Gerard. (ed.): *Minority Peoples in the Age of Nation States*. Translated by T. Berrett, London: Pluto Press, pp. 119–125.

MARTIN, Ruth. 2007. *Overseas Transfer and the Japanese Housewife: Adapting to Change of Status and Culture*. Folkestone: Global Oriental.

MORGAN, Glenn et al. 2003. "Global Managers and Japanese Multinationals: Internationalisation and Management in Japanese Financial Institutions". In: *International Journal of Human Resource Management* 14 (3): pp. 1–19.

NEWHALL, Paul. 1996. *Japan and the City of London*. London: Athlone.

PUTNAM, Robert D. 1995. "Bowling Alone: America's Declining Social Capital." *Journal of Democracy* 6 (1): pp. 65–78.

SAKAI, Junko. 2000. *Japanese Bankers in the City of London: Language, Culture and Identity in the Japanese Diaspora*. London, New York: Routledge.

TURNER, Edith. 2012. *Communitas: The Anthropology of Collective Joy*. New York: Palgrave Macmillan.

TURNER, Victor W. 1969. *The Ritual Process: Structure and Anti-Structure*. New Brunswick, London: Aldine Transactions.

YOSHIDA, Miyako. 2011. "Former Principal Dancer, Royal Ballet play4jpn". YouTube video, 3:57. Posted 28.3.2011. https://www.youtube.com/watch?v=iedifbFncB4. (accessed October 17, 2012)

The Triple Disaster as an Opportunity to Feel Japanese Again in Hawaii

Jutta Teuwsen

1 Introduction

This paper will analyze the effect of the 3/11 Triple Disaster on the process of identity formation of Japanese Americans living in Hawaii. Of all Japanese diasporas, Hawaii's has a special significance because in relation to the overall population, its Japanese American population is larger than all other Japanese diasporas abroad. Historically, Hawaii was one of the first destinations to where large groups of Japanese workers migrated during the late nineteenth century.

During a fieldwork project focusing on the Japanese department store Shirokiya between August and November 2011, I realized that the construction of a specific Japanese identity in Hawaii was in a period of transition and that 3/11 played an important part in this ongoing process. In interviews conducted with nisei and sansei, it became evident that not only their feelings of identity and their understanding of and dealing with their Japanese heritage was influenced by this overseas catastrophe, but also the social conditions for Japanese Americans in Hawaii have been changing since the disaster.

In order to understand what effect 3/11 had on the status of the Japanese Americans in Hawaii and how these changes can be explained, this paper will give attention to five aspects. First, I will give a brief overview of the Japanese American population in Hawaii as it is today. Second, I will focus on the history of Japanese immigration to Hawaii, followed by an analysis of the social status of the Japanese American diaspora and then the reaction of the Hawaiian population on 3/11. This paper ends with a short outlook on the future of Japanese Americans in Hawaii after 2011.

2 Japanese American Hawaii

According to the 2010 US census, nearly 23 percent of Hawaii's population claimed to be Japanese or Japanese in combination with at least one other race.

Of these, more than 185,000 – almost two-thirds – claimed to be only Japanese, which is 13.69 percent of Hawaii's total population. It is striking that even after five generations, *gosei*[1] still claim to be Japanese. Even now, the majority of nisei does not speak Japanese fluently, perhaps because their parents' generation, the issei, had to endure growing nationalistic sentiments in the US after World War I, and as a result learned to adjust to the English language (Kimura 1988: 185). Therefore, even if their parents were native Japanese speakers, subsequent generations spoke Pidgin English, which evolved as a major language on the plantations where the majority of Japanese immigrant laborers worked. For the nisei generation, it meant that they grew up in an English-language environment; thus, English became their first language. If these second generation Japanese studied Japanese at all, they learned it as a second language in Japanese language schools. However, after the Japanese attack on Pearl Harbor on December 7, 1941, the Japanese language schools closed down and the public use of Japanese language was even prohibited because, in a socio-political context, Americans of Japanese descent were suspected to be secretly working for their mother country (Kimura 1988: 225; Odo 2003: 101–116). Only in 1947 did some of the Japanese language schools reopen, but by then the nisei generation had hardly spoken Japanese for many years (Kimura 1988: 254). Thus, while the majority of the nisei might have or have had only basic knowledge of the Japanese language, the following generations generally do not speak or understand Japanese.

The situation for the Japanese Americans in Hawaii, therefore, differs significantly from the typical transient Japanese experience elsewhere in the world. Many Japanese live only temporarily in London, Brussels, Dusseldorf and Paris. They are dispatched by their companies and accompanied by their families but will return to Japan after a few years abroad. In contrast, people of Japanese origin in North and South America and in Hawaii can often trace the moment of their families' immigration back four or five generations. Accordingly, they see themselves as Americans with a Japanese background. Many are born in the US with English as their native language. Japan is certainly a popular destination for

---

[1] Gosei are fifth-generation Japanese.

tourists and a place that is somehow connected to nostalgic feelings for Hawaiian Japanese Americans, but they still consider their homeland to be Hawaii – that is, the US. However, there is also a minority that rather fits to the description of a transient diasporic community. This group consists of individuals coming to Hawaii for a certain period of time, going back to Japan after one or two years. In general they are surfing teachers or occupy other jobs related to Hawaii's tourist industry that significantly depends on Japanese tourists. However, the majority of individuals with Japanese background are descendants of the Japanese immigrants from around 1900.

3   History of Japanese Immigration to Hawaii

In 1868, year one of the Meiji-period, the first group of 148 Japanese men went to Hawaii to work on the sugar plantations. These *Gannen-mono* (Meiji first-year people) mark the beginning of Japanese diaspora (Befu 2001: 3–22; Danniels 2006: 31). The Japanese workers came to Hawaii on a three-year contract. Initially, all migrant workers intended to return home after having earned sufficient money. They had the plan to take the money back home to Japan and have a better life with their families (Befu 2010: 34). However, due to bad working conditions, some of the workers went home even before their contracts ended, and only 90 gannen-mono decided to settle down in Hawaii (Kimura 1988: 3). After successful negotiations regarding changes to working conditions and the improvement of contracts, Japanese immigration increased and more migrant workers settled down in Hawaii and stayed for good. By 1898, more than 16,000 Japanese were living in Hawaii (Kimura 1988: 13). In the same year, Hawaii became a territory of the US, this had significant effects on the Japanese immigrants to Hawaii. As one consequence, Japanese no longer had to find contracts with sugar plantations in order to be allowed to migrate to Hawaii. Between 1900 and 1907, more than 68,000 Japanese went to Hawaii; not all of them stayed, and more than 35,000 Japanese left for the west coast of the US, drawn by the expectation of higher wages (Kimura 1988: 13–15). Due to the high numbers of Japanese immigrants, an informal "Gentlemen's Agreement" was arranged between the US and Japan in 1907/08 (Sawada 1991: 339–359). From that time forward, Japan restricted passports, issuing them only to close kin and "picture brides" of those Japanese already living in the US.[2] Finally, in 1924, a clause of the US Immigration Act completely barred Japanese immigration and led to strong tensions between the Japanese government and the US and also sparked weak movement albeit Japanese Americans aimed for the revision of the Act (Hirobe 2002). In that year, 125,368 Japanese were living in Hawaii. After immigration was legalized again, Hawaii returned to its status as a popular destination for Japanese immigrants. Even today, the Japanese American population remains one of the largest "racial" groups in Hawaii according to the 2010 census (US Census Bureau 2010).

4 The Social Status of Japanese Americans in Hawaii

The outlined historical circumstances indicate why we cannot compare the Japanese American diaspora of Hawaii with transient diasporas in Europe. The Japanese are an integral part of the overall population of Hawaii and see themselves as locals. The term "local" is to some extent opposed to what is generally understood as "diaspora": "local" represents the shared identity of all Hawaiian people who have an appreciation and commitment to the islands, people and way of life. Being local in Hawaii is inclusive – local people can be of Japanese, Chinese, Portuguese or any other ancestry. They are local first and foremost. Finally, being local is not about one's heritage but about presence; it is about living in Hawaii right now, which suggests that Hawaii's Japanese Americans deal with their history of immigration in a different way compared to other Japanese diasporic communities. However, there are also historical reasons why this group does not publicly demonstrate its Japanese background. Over time, the image of the Japanese American population changed continuously.

In 1909 and 1920, the Japanese on the sugar plantations went on strike for better working conditions. The Filipino Labor Union initiated the second strike in 1920 (Duus 1999). However, the Japanese workers joined belatedly, and union

---

[2] Picture brides were Japanese women sent from Japan to Hawaii in order to enter into an arranged marriage. The practice got started when the immigrants realized that they could not accumulate enough money to go back to Japan soon trough their work (Yun Chai 1988). The future husbands had nothing but a picture, waiting for the brides to arrive at the harbor.

leaders on both sides mistrusted each other and diverged on the general aims of the strike (Pierce 2007: 577). These strikes were only partly successful and lead to a deep resentment towards the Japanese by groups of workers. As a result, many Japanese left the cane fields. The number of Japanese who worked on the fields declined only from 1930 onwards, and those who remained increasingly occupied minor positions of responsibility, like field overseers (Kimura 1988: 100f.). As a result, both working and living conditions improved noticeably, and the nisei born around 1930 did not experience the constraints of the earlier years.

While the conditions on the cane fields improved and increasing numbers of issei advanced to positions with more responsibility, a solid base for the following Japanese American generation was built: major institutions and organizations had been established and took root, like the United Japanese Society in 1932, and Japanese vernacular newspapers were widely spread (Kimura 1988: 178f.) The political, economic and social influence of the Japanese American population in Hawaii grew and more and more parents could afford to send their children to colleges on the US mainland. The results are still visible today – as a group, Japanese Americans are socioeconomically very successful (Okamura 2008: 127).

Japanese Americans in Hawaii occupy influential administrative, professional and clerical occupations on all state levels and are highly visible. However, from the 1970s onwards, this visibility gradually evolved into a negative stereotype and the Japanese American communities were resented because of their success. Furthermore, the Japanese became the scapegoats for social and economic problems. As a consequence, this group refrained from stressing their Japanese identity and instead focused on their local identity.

Because the Japanese American community tended to hold back their ethnic background, it is promising to analyze the groups' reaction following the triple disaster. During my field research in Hawaii, I interviewed a number of Japanese Americans, especially those with ties to the prefecture of Fukushima and also approached the Japanese Cultural Center of Hawaii and the Ethnic Studies Department of the University of Hawaii. The impression I got from my interviews seemed to be in line with the opinions of the experts: The "public denial"

of a Japanese heritage did neither result in a much less visible reaction by the
Japanese communities towards 3/11 compared to other diasporic communities
nor in a highly visible reaction. Their reaction did not differ from other Japanese
American groups in Hawaii. And they did not play a key role in promoting the
visibility of the Japanese American communities after 3/11. To sum it up, the reaction towards 3/11 was intense through all social strata and local communities
in Hawaii and unrelated to ethnic background or geographical descent in Japan.

5   Reactions On 3/11

Following the 3/11 catastrophe, there have been beneficial events and donations
as well as other activities. Furthermore, the overall presence of the incident –
as well as the presence of Japanese Americans – in the media has been striking. It is remarkable to what extent the triple disaster dominates the economical
and political sections in Hawaii's local newspapers from 2011. In that respect,
Hawaii is probably not different from other places because people all over the
world showed their interest and compassion for Japan and the Japanese. However, the triple catastrophe also dominated local news. The Japanese Cultural
Center of Hawaii initiated many benefit events, and the Japan-America Society
of Hawaii started a great part of the fund raising. Likewise, local organizations
were and are still engaged in aiding Japan. The Aloha for Japan initiative involves more than 160 Hawaiian financial institutions, merchants, schools and
community groups. Because Hawaii depends significantly on Japanese tourism,
companies like Hawaiian Airlines and major hotel groups also felt the need to
show their support through financial donations for aid and reconstruction.

Many Hawaiians participated in events and donated individually. One example of an event is the Lei Day for Japan. A *lei* is a Hawaiian necklace made of
flowers. Respected individuals and guests are honored by receiving a *lei*, which
is a crucial element in many festivities. The *lei* is strongly connected to the concept of friendship and is supposed to foster feelings of belonging. Since 1928, Lei
Day has been a national holiday, and Steven J. Friesen has pointed out how the
lei as well as Lei Days have contributed to the construction of an ethnic identity
on the Hawaiian islands (Friesen 1996: 1–36). This festival is celebrated on May

1, and in 2011 it was turned into the Lei Day for Japan by Hawaian Airlines. The airline organized a music and cuisine benefit event held at the very prominent Aloha Tower Marketplace in Honolulu; the benefit raised more than $ 200,000.

Finally, the numbers provided by the American Red Cross Hawaii State Chapter are most telling: The people of Hawaii, with a population of not more than about 1.3 million people, raised over six million dollars to aid the Japan relief effort. Many schools, NGOs, companies and individuals made donations ranging from $ 1 to $ 50,000. Examples of fund-raising activities include car washes, bake sales and a telethon with a concert held on the lawn of the Hilton Hawaiian Village; here performers donated their time and talent while people from around the world watched on television and/or via computers and made donations to the Red Cross online as well as by phone. A group of companies and banks worked together with the Hawaiian Red Cross and Japan American Society of Hawaii to coordinate efforts on all islands.

I would like to stress that the Hawaiian population donated significantly higher amounts of money to the victims of 3/11 than they had ever donated before for other catastrophes like Hurricane Katrina or 9/11. During the first two weeks after the earthquake, the American Red Cross Hawaii State Chapter received more than $ 640,000 from Hawaiian residents, which was more than triple the $ 190,000 sum that was donated two weeks after the Haiti earthquake.

The generosity of the Hawaiian population can be traced back to four reasons. The first reason is that both islands, Hawaii and Japan, were threatened by the same tsunami. It hit Japan worst, but the Hawaiian people are highly aware of the fact that it could have happened to them. This feeling was intensified by a historical precedence: in 1960, a massive earthquake close to the coast of Chile (magnitude 9.5), which was actually the strongest recorded earthquake in the twentieth century, led to a tsunami that traveled around half the globe. It first hit Hawaii, after crossing 10,000 km of the Pacific, and killed 61 people. However, the tsunami did not stop in Hawaii and reached the coast of Tōhoku, the site of the triple catastrophe in 2011, eight hours later. Government officials as well as the general population in Japan did not expect that the tsunami would still be powerful enough to cause damage when reaching the shore and were thus unprepared. Although the news had broadcast that Hawaii had been hit by a tsunami, no official warning was released for the Eastern Coast of Japan (Yoshimura 2004: 156). When the tsunami finally reached Japanese shores, it killed another 161 people.

Thus, Hawaiian people in 2011 took a highly empathetic position because they could easily imagine that they could have experienced a similar scenario. Nevertheless, the tsunami of 2011 caused $ 22 million in property damage in Hawaii and forced the closure of two hotels. For Japanese Americans, being reminded of the Great Hanshin earthquake of 1995 might even have led to a flashback experience because they felt personally involved through their Japanese heritage, and as a result, they reacted with even more empathy.

The second, and probably more influential reason for the high amount of donations given by the Hawaiian people is that the Japanese Americans constitute such a high percentage of the overall population, which means that they are present and influence or affect all socio-economic spheres. This presence of the Japanese Americans, although they refrained from publicly showcasing their Japanese American identity, is still highly visible all over the island. There are Japanese restaurants, and green tea drinks and *matcha* ice cream (green tea ice cream) are available throughout the islands. The major Japanese department store Shirokiya defends a very prominent spot at the country's biggest shopping center: the Ala Moana Center in Honolulu. It was to be closed several times, but different interest groups are still fighting to keep it open. The crucial aspect is that things "Japanese" are considered to be a significant characteristic of Hawaii for all local people – independent from their ethnic background. Even if the perception of the Japanese Americans might be judged positively or negatively, they are in fact a part of the locals' everyday life. Thus, 3/11 affected not only the Japanese American communities but also each and every individual claiming to be local.

Third, the constant and visible presence of tourists from Japan also ought to be seen as a significant reason for the contributions of local communities. Most of the local communities are aware of the fact that Japanese tourism is one of the major pillars of the prosperous local economy. Of all tourists, 17.3 percent who came by air in 2011 came from Japan. Nearly two million Japanese fly to Hawaii every year. Japanese tourists spend the most per day, averaging $ 289 whereas the average amount of money spent per day by tourists is only $ 179. Contributions to the relief efforts are therefore important to reciprocate the Japanese tourists' generosity. Furthermore, the donations make sure that the positive image Japan has of Hawaii will remain at the same high level and uphold a continuous stream of Japanese tourism to Hawaii. It is certainly not by accident that Hawaiian Airlines was keen to organize the Lei Day for Japan.

Finally, the high amount of individual funding can be traced back to the long tradition of gift giving and reciprocity of local groups in Hawaii. Colleen Leahy Johnson analyzes the reasons for this practice in the case of Japanese Americans in Honolulu (Johnson 1974: 295–308). She claims that gift giving aims at keeping a balanced reciprocity, serving all essential social functions. In detail, she differentiates four means of gift giving to support and promote social functions: "facilitating status placement, providing general continuities, reducing conflict, and equalizing class difference" (Johnson 1974: 295). The reciprocity of individual funding after 3/11 might point to an acknowledgement of what Japanese Hawaiians and tourists have done for Hawaii. Donors certainly could not have expected direct reciprocal returns for their support after 3/11.

6 The Future of Japanese Americans in Hawaii after 2011

As argued earlier, the status of the Japanese American community has changed during the last decades. Japanese Americans did not promote their ethnic identity actively because they were resented due to their success. They were used to being made responsible for social and economic problems, but conditions have significantly changed since the 3/11 catastrophe. The negative perceptions of the Japanese American communities by other locals appeared to be obsolete because the triple disaster was the first thing that came on the locals' mind in the context of Japan or anything Japanese. A complex construct of prejudice, experience and shared history with the Japanese Americans had evolved over more than one decade, but it seems that it was reset in one day. After decades of holding back their Japanese heritage, 3/11 allowed Japanese Americans to show their heritage freely and openly again. Apart from their local immersion, now they are again placed in the context of the homeland of their acclaimed ancestors.

To conclude, 3/11 has brought about significant changes in the self-image as well as in the perception of the Japanese American communities of Hawaii. It will be interesting to see how this new setting will influence the future image and self-image of the Japanese American groups. As time goes by, former and well-known structures of behavior and perception might find their way back; nevertheless, there is a chance that the new setting might persist.

Literature

AMERICAN ANTHROPOLOGICAL ASSOCIATION. 1997. "Race and Ethnic Standards for Federal Statistics and Administrative Reporting". Response to OMB Directive 15. http://www.aaanet.org/gvt/ombsumm.htm. (accessed 3 January 30, 2012)

BEFU, Harumi. 2001. "The Global Context of Japan Outside Japan." In: Befu, Harumi; Guichard-Anguis, Sylvie (eds.): *Globalizing Japan: Ethnography of the Japanese Presence in Asia, Europe and America.* London: Routledge, pp. 3–22.

BEFU, Harumi. 2010. "Japanese Transnational Migration in Time and Space: A Historical Overview." In: Adachi, Nobuko (ed.): *Japanese and Nikkei at Home and Abroad. Negotiating Identities in a Global World.* New York: Cambria, pp. 31–49.

DANIELS, Roger. 2006. "The Japanese Diaspora in the New World: Its Asian Predecessors and Origins." In: Adachi, Nobuko (ed.): *Japanese Diasporas: Unsung Pasts, Conflicting Presents, and Uncertain Futures.* London: Routledge, pp. 25–34.

DUUS, Masayo. 1999. *The Japanese Conspiracy: The Oahu Sugar Strike of 1920.* Oakland, CA: U of California P.

FRIESEN, Steven J. 1996. "The Origins of Lei Day: Festivity and the Construction of Ethnicity in the Territory of Hawaii." *History and Anthropology.* 10 (1): pp. 1–36.

HIROBE, Izumi. 2002. *Japanese Pride, American Prejudice: Modifying the Exclusion Clause of the 1924 Immigration Act.* Stanford: Stanford UP.

JOHNSON, Colleen Leahy. 1974. "Gift Giving and Reciprocity Among the Japanese Americans in Honolulu." *American Ethnologist.* 1 (2): pp. 295–308.

KIMURA, Yukiko. 1988. *Issei. Japanese Immigrants in Hawaii.* Honolulu: U of Hawaii P.

ODO, Franklin S. 2003. *No Sword to Bury: Japanese Americans in Hawai'i During World War II.* Philadelphia: Temple UP.

OKAMURA, Jonathan Y. 2008. *Ethnicity and Inequality in Hawai'i.* Philadelphia: Temple University Press.

PIERCE, Lori. 2007. "Hawaii Laborers' Association." In: ARNESEN, Eric (ed.) *Encyclopedia of U.S. Labor and Working-Class History.* New York: Routledge, pp. 576–577.

SAWADA, Mitziko. 1991. "Culprits and Gentlemen: Meiji Japan's Restrictions of Emigrants to the United States, 1891–1909." In: *Pacific Historical Review.* 60 (3): pp. 339–359.

U.S. CENSUS BUREAU. 2010. http://www.uscensus2010data.com/. (accessed January 29, 2012)

YOSHIMURA, Akira. 2004. *Sanriku kaigan ōtsunami [Great Tsunamis on the Coast of Sanriku].* Tokyo: Bungei shunjū.

YUN CHAI, Alice. 1988. "Women's History in Public: 'Picture Brides' of Hawaii." In: *Women's Studies Quarterly.* 16 (1/2): pp. 51–62.

Disaster, Donations, and Diaspora: The Response of the Japanese-Brazilian Community of São Paulo to the Triple Disaster of 2011

Peter Bernardi

The triple disaster or 3/11, the devastating combination of earthquake, tsunami and the meltdown of reactors at the nuclear power plant Fukushima Daiichi that happened in Japan on March 11, 2011, was a globally reported catastrophe. The Japanese diasporic communities especially demonstrated their willingness to help, organize campaigns and send donations. While a commitment to restoring and relating to the homeland in some way confirms William Safran's (1991: 84) initial criteria for diasporas, the study of the Japanese diasporic communities' response to 3/11 also offers insights into these communities and their status. This paper offers an analysis of the impact of 3/11 drawn from the reaction of the Japanese-Brazilian diaspora in São Paulo. Although initiatives to help Japan were started throughout Brazil, São Paulo hosts the largest and most influential *nikkei* community that organized the biggest campaigns. Furthermore, the article is my attempt to relate observed connections between disaster, donations and diaspora. I had originally scheduled fieldwork on Japanese immigration for March 2011, but after arriving in Brazil on March 12, reactions to the catastrophes of 3/11 were a constant topic in many encounters and overlapped my intended research on the *nikkei* community of São Paulo. Donations played an important role, and in this article I argue that help (especially financial) after 3/11 was not only about Japan but also the self-perception of the *nikkei* community in São Paulo. By donating and participating in the campaigns, Japanese-Brazilians were able to perform their identity as "Japanese" for the community itself, for the Brazilian society, and also for representatives of official Japan. To provide some background, a short overview over the *nikkei* community in São Paulo is followed by a historic focus on two previous initiatives for helping Japan. After reporting about the general response in Brazil to the catastrophes, the article then concentrates on São Paulo and presents how main actors reacted, how campaigns were perceived in the community, and which implications can be deduced.

1 Japanese Diaspora in Brazil

Brazil ranks first among the worldwide Japanese diasporas, counting more than 1.5 million Japanese and those with a Japanese heritage. Historically, Japanese immigrants were sent to Brazil in 1908 in order to work on coffee plantations and balance labor shortages after the abolishment of slavery in the late 19th century. Economic success came through individual and cooperative farming, along with a transition to urban areas and white-collar jobs. Today, many descendants of these settlers have gained considerable economic wealth and are seen as a "respected and highly educated immigrant group" (Brody 2002: 49). A disproportionally number of *nikkei* are not only students but also educators at schools and universities (Adachi 2004: 63). They have achieved political success as city councilmen, delegates and even ministers in different political administrations (Harada 2010). Among scholars, Japanese immigrants to Brazil and their descendants are therefore considered to be a "positive" or even highly integrated "model minority" (Adachi 2006: 12). In Brazilian society, the *nikkei* retain a positive (yet stereotypical) image and are often referred to as "Japanese" whose continued connection to Japan, their former homeland, is strongly assumed (Lesser 2007: 150). The *nikkei*'s relationship with Japan has become more contentious since the 1990s, when the Japanese government changed immigration policies to attract foreign workers with Japanese ancestry. Especially Brazilian *nikkei* have gone to Japan to work in mainly blue-collar jobs and have settled there, even though they experience a stark contrast of their perception as undesirable foreigners. These workers-away-from-home or *dekasegi* now form the country's third largest minority, suffering from both discrimination and the effects of a declining economy (Reis 2002; Brody 2002).

In Brazil, a majority of *nikkei* live and work in the state of São Paulo and its capital, the city of São Paulo, where they have formed a visible community. The district of Liberdade is known as the "Oriental" or "Japanese district" because many

Japanese immigrants once lived and established businesses here.[1] Liberdade is also home to the main actors in the *nikkei* community of São Paulo, e. g.: its central entity, the association Sociedade Brasileira de Cultura Japonesa e de Assistência Social (*burajiru nihon bunka fukushi kyōkai*, abbreviated as "Bunkyo"), most prefectural associations (*kenjinkai*), and various other organizations. Apart from these official institutions, the daily Japanese-language newspapers *São Paulo Shimbun* and *Nikkey Shimbun* and their Portuguese counterpart, the weekly *Jornal Nippak*, a Buddhist temple, restaurants serving sushi, sashimi and ramen, Japanese convenience and specialty stores, and even banks attending customers in Japanese can be found in Liberdade.

With this concentration of formal and informal venues, Liberdade was a center of action and reaction in São Paulo after the catastrophe of March 2011. In the main plaza next to a bank styled after a medieval Japanese castle, gatherings were held and donations collected. Newspaper reports were posted in the offices of various *kenjinkai* where people from the prefectures met to communicate, making these offices hubs for gathering and distributing information as well as for coordinating help for Japan. Private initiatives appeared throughout the district and offered opportunities to voice feelings of grief and solidarity. One example often referred to in the media was the café Kōhī whose *nikkei* staff started a campaign not only to collect money but also to provide moral support. Guests were asked to leave messages or sign a twelve-meter long poster afterwards sent to Japan in cooperation with the *dekasegi* of Miyagi prefecture. Social media networks such as Orkut, the Brazilian equivalent to Facebook, were used to gather support and promote campaigns like Gambare Japão whose aim was to produce video messages in which Brazilians and Japanese-Brazilians alike expressed their wishes for a speedy recovery, ending with the phrase Gambare Japão. Other examples included a mass with the participation of Protestant, Catholic and Buddhist priests in the central auditorium of Bunkyo, fund-raising events such as dinners organized by Japanese-Brazilian chefs in São Paulo's city hall, concerts by the state's symphony orchestra, and art expositions. Labeling of these events ei-

---

[1] Even though this name refers the visible presence of both Japanese, Chinese and Korean immigrants and their descendants, nowadays, the latter form a majority of the residents.

ther gave mixed-language statements in Japanese and Portuguese, such as Gambare Japão and Nippon Ganbare, or implicitly called for help with titles such as
SOS Japão or Everyone United for Japan. This multitude of campaigns and events
show that the catastrophes of 3/11 were very visible moments. Various actors organized the help for Japan and the various campaigns were not only reported in
the media specifically used and produced by Japanese and Japanese-Brazilians
but also gained attention in other newspapers and television programs. However, the fact that the Japanese diaspora of Brazil helped Japan after a natural
disaster is by no means a new phenomenon. By linking the events of 2011 to
other occasions, conclusions can be drawn about the relationship between Japan
and Japanese descendants abroad.

2 Helping Japan

The relationship to Japan, especially in times of crisis, was always a topic among
Brazilian *nikkei*. One of the earliest efforts took place in 1947. During the Second World War, the Brazilian state sided with the Allied Powers and especially
discriminated against Japanese – as well as German and Italian – immigrants as
a suspected fifth column. After Japan's surrender, the Japanese community in
Brazil was sharply divided between two groups. Those who believed that Japan
had actually won the war (*kachigumi*) defended their view, sometimes with organized terrorist attacks, against the minority that accepted the defeat (*makegumi*)
(Adachi 2004: 59). Only after the Brazilian state and Japan finally intervened in
the early 1950s was this situation resolved – but at the cost of a confused community. In 1947, a group of Japanese immigrants in São Paulo founded the Committee to Help the Victims of the War in Japan (Comitê de Socorro ás Vítimas da
Guerra no Japão) and organized donations to post-war Japan. The committee
acted with the support of the Brazilian Red Cross and worked together with the
American occupiers, sending goods and CARE packages worth about $15,000
to Japan until July 1950 (Handa 1987: 743). Apart from providing help to the
immigrants' homeland, the committee was also considered to be an institution
that might help to unite the divided immigrants through a common cause. But
the *makegumi* group's cooperation with the Americans, a former enemy, intensi-

fied the conflict between *kachigumi* and *makegumi* groups. Because only a small number of the immigrants participated and the project met massive resistance from the *kachigumi* groups, the campaign failed in this larger goal of uniting the immigrants (Beltrão 2008: 257).

The last event that involved Brazilian *nikkei* in efforts to help Japan was the earthquake of 1995 that devastated Kōbe city and southern parts of Hyōgo prefecture. It claimed about six thousand lives, the majority of them inhabitants of Kōbe. With an estimated five thousand Brazilian *nikkei* living in Hyōgo prefecture and about one thousand in Kobe itself, the community in Brazil tried desperately to contact friends or family in Japan (Takezawa 2002: 313). Uncertainty continued with rarely-working telephone lines and delayed official information until the Japanese government confirmed the first Brazilian *nikkei* victim. Of 173 foreign victims counted, eight were Brazilian *nikkei* (Takezawa 2002: 315). The *nikkei* media in Brazil widely reported their fates and also the stories of those who were injured or had lost their belongings or housing (São Paulo Shimbun 1995a). These very personal links to the disaster directly involved and affected the community. Various actors as the *kenjinkai* of Hyōgo province or Bunkyo met, and although they started individual campaigns for material and moral support, no organization assumed responsibility for a joint campaign (São Paulo Shimbun 1995b). Financial donations were individually collected and likewise distributed; the Hyōgo *dekasegi* transferred its donations directly to the government of Hyōgo prefecture itself (São Paulo Shimbun 1995c: 1). This individualization of donations was also present in the campaign of Banespa Bank, one of the biggest Brazilian banks with branches in Japan. Banespa started the S.O.S. Japão campaign, the donations of which were to be used to support Brazilian NGOs that helped especially Brazilian *dekasegi* and Brazilian victims of the earthquake (São Paulo Shimbun 1995d: 4). In total, Brazilian *nikkei* contributions to the affected prefectures were estimated at $ 820,000 (Takezawa 2002: 316).

We can consider both reactions in 1947 and 1995 to be a form of solidarity in which the group that received help was clearly identified as one's own. Helping in 1947 meant helping a Japan that the immigrants felt an allegiance to and had until recently called their homeland. In 1995, this relationship had changed and help also targeted various compatriots from Brazil and those who were, in turn, helping them. The example of Kobe displays a more heterogeneous direction of help. The catastrophes of 3/11 show different perspectives on help from a diasporic community.

3   Reactions to 3/11 in Brazil

The Japanese government requested and received international assistance after the catastrophes of 3/11. Countries such as Australia, the United Kingdom and South Korea dispatched earthquake rescue teams. Others sent relief goods, food, blankets, and emergency lamps, with the majority of goods coming from China and the United States of America (Okada et al. 2011: 39). The Japanese government of Prime Minister Naoto Kan also received various offers of further help and decided that all financial help after the catastrophe was to be given to the Japanese Red Cross. Worldwide, more than one billon dollars were collected by April 3 (Okada et al. 2011: 39). Fund-raising was therefore a strong focus in many diasporic communities.

On a national level in Brazil, President Dilma Rousseff immediately sent an official note to Prime Minister Kan. While President Rousseff expressed her condolences and Brazil's readiness to participate in international efforts to help Japan, she also specifically mentioned the special connection to the Brazilian nationals living in Japan (Blog do Planalto 2011). As one of the first officials, Brazilian Minister of External Relations Antonio Patriota called his Japanese colleague Takeaki Matsumoto to express his condolences and Brazil's solidarity. The Brazilian government followed with a donation of $ 500,000, which it transferred to the Japanese Red Cross, conforming to the wishes of the Japanese government (São Paulo Shimbun 2011a).

Reactions among the *nikkei* community in São Paulo can be divided into two phases: a shorter state of shock and confusion followed by an active and longer phase of organizing concrete help. After Brazilian television broadcasted the pictures of earthquake and tsunami, many tried to contact friends or family members working in Japan. As one of my contacts in the field mentioned, the Brazilian media in these first days visited many associations of the community and reported emotional stories of missing relatives and their fates. But these stories soon diminished because in stark contrast to the catastrophe of Kobe in 1995, no Brazilian *nikkei* lost his or her life during 3/11.[2]

A quote from a representative of a *nikkei* NGO in the weekly *Jornal Nippak* sums up the second phase in a nutshell: "What the Japanese need now is moral support and financial help" (Shiguti 2011a).[3] Immediately after the earthquake, leaders of influential Japanese-Brazilian associations met in Liberdade to coordinate a joint response. The aforementioned Bunkyo, the federation of the prefectural associations (*kenren*), the philanthropical Beneficência Nipo-Brasileira de São Paulo (Enkyo), the cultural association Aliança Cultural Brasil-Japão, and the Chamber of Japanese Commerce and Industry in Brazil decided to launch a combined campaign to collect donations; the campaign was formally named the Campaign to Collect for the Victims of the Earthquake in Japan. The organizers created four different bank accounts for donations. In the beginning, they did not decide exactly who should receive these donations, but on March 22, the committee announced that all donations would be transferred to the Japanese Red Cross, adding that city- and state-wide responses were promising (Abe-Oi 2011c). The campaign raised more than 3.2 million Reais, about one million dollars, in the initial ten weeks (Abe-Oi 2011e). In addition to participation from actors in São Paulo city, many Japanese associations in the interior of São Paulo state submitted substantial donations ranging from about $ 1,000 to $ 5,000. The campaign also counted small donations from unidentified donors, supposedly Brazilians without Japanese ancestry. This was a huge effort for the community and its members. Bunkyo was the organization that made this process public by presenting donors and their reasons to participate on their homepage. The organizing committee also communicated the amount of donations in weekly press conferences, although they had agreed not to communicate the amount of each one's collection.

---

[2] Interestingly enough, media in Brazil concentrated more on the dimensions of the natural disasters of earthquake and tsunami while in contrast, German media reporting put its main focus on the efforts to control the nuclear power plant.

[3] All translations of Portuguese sources are from the author.

The keyword in this campaign was "solidarity" (*solidariedade*), but its interpretations were ambiguous. As an idea, solidarity in Brazil has not only been
linked to Christian and socialist ideals but also was fixed in 1988 as a fundamental principle of Brazil's constitution (Rosso 2007: 206). Definitions of the word
connect it to a humanitarian idea of helping "those unprotected, those who suffer, the unjust treated" with an "intent to comfort, console and offer help" yet
limits it to a specific group that shares "common attitudes and sentiments [...] to
form a strong unit" (Houaiss 2001: 2602). In the early immigrant *nikkei* community, solidarity was also strong principle for mutual help (Handa 1987: 771).
This complex interpretation manifested in the "messages of solidarity" released by the organizing committee to different audiences: to the Japanese government of Prime Minister Kan and to the *nikkei* in Japan and Brazil. All messages published on Bunkyo's homepage drew heavily on the term "solidarity"
and expressed the intent to help, but their focuses were different. The message
in Portuguese and Japanese addressed Prime Minister Kan "in the name of the
Japanese-Brazilian community" and established the status of the signees as representatives of the community who acknowledge and praise political initiatives
of Kan's government. The organizing committee announced that the community
"was studying possible ways of collaborating with the efforts, using all of its modest capacity" (Abe Oi 2011b) but gave no further details although bank accounts
for receiving donations were already set up. Solidarity in this context meant po-
sitioning oneself in a political context. The messages to Brazilian *nikkei* in Japan
and Brazil were more practical. They asked them to remain calm, inform their
relatives in Brazil when possible, and also announced the start of the campaign
"with the aim to find a form to show our solidarity to Japan" (Abe Oi 2011a).
These also invoked a sense of a larger (Japanese) community, assuring that "we
want to express profound solidarity because we share the great pain that over-
whelms the whole Japanese nation" (Abe Oi 2011a). These communiqués once
again established the organizers as responsible representatives of the Japanese-
Brazilian community, whether in Japan or Brazil. But by focusing on solidarity,
the organizers included the *nikkei* of Brazil in a concept of Japanese community

that surpassed national borders. This idea of a solidarity among Japanese is also apparent in the way donations were given and received.

A donor could theoretically just transfer the money to the bank accounts created for this occasion. This procedure would be much easier, more transparent and even safer for all parties concerned. However, many donations from the *nikkei* community were made in person, and especially those coming from Japanese associations, clubs and circles (*nihonjinkai*) followed a very similar pattern; when those donors came to the Bunkyo building in Liberdade to deliver their contribution, they were received by Bunkyo's president, Kihatiro Kita, or a colleague from the administration. As the donor formally delivered the check, the Bunkyo staff took pictures and posted a report online afterwards. The donors in this situation presented themselves as participants and expressed their (expected) solidarity with Japan, further confirmed by pictures, reports and donation receipts. Donating money became a performative act in which members of this group confirmed their right to belong. Solidarity in this sense was shown and proven.

4 Reacting to 3/11: Solidarity with the Japanese

The donors to the general campaign to help Japan ranged from individuals and Japanese associations from the interior of São Paulo state to sport clubs, cultural institutions and *nikkei*-owned companies. The Bunkyo homepage functioned as an official forum that presented these contributors and their reasons to participate. Especially among smaller contributors, help for Japan was explained as help to a group they considered themselves to be a part of, the group of "Japanese". This idea of solidarity expands from a humanitarian idea to a very specific feeling of solidarity within a group. The opposite of Gayatri Spivak's concept of "othering" (Spivak 1985), this process might be called "ouring". Participating in the campaign meant demonstrating one's own link to Japan and identifying as Japanese among the community in Brazil.

Peter Bernardi

The association Shimbokukai e Fujinkai de Vila Santa Izabel donated the relatively small sum of about $ 1,500.[4] In the interview on the Bunkyo homepage,
the three female representatives of the Fujinkai framed their participation as a
natural act: "We are very grateful to Japan, and this is the moment to return the
favor to the country our ancestors came from and where our children now live"
(Abe-Oi 2011d). References to family ties run through the whole interview: to
help Japan meant to help "our Japanese brothers and sisters". The speakers define themselves as members of this Japanese family that includes Brazilian *nikkei*
in Japan (Abe-Oi 2011d). In contrast to larger contributions that resulted from
refined campaigning, the example of the Fujinkai also shows participation on a
smaller scale. The Fujinkai is located in southern São Paulo and operates in a
modest neighborhood with few families of Japanese ancestry. Its members are
mostly elderly Nisei and the association functions as a contact point for their
informal activities, such as karaoke, Japanese language classes, *rajio taisō* (radio
gymnastics) and bingo games. Aside from the actual amount of the contribution,
there also exists an emotional meaning. The money came not from a specific
campaign but from the personal savings account of the Fujinkai. The contribution was thus understood as a sort of familial legacy. The members of Fujinkai
explained that its money should be better used to help Japan since their "association will not survive for much longer" because of their own ages and missing
younger members (Abe-Oi 2011d).

In a way, the example of the Fujinkai symbolizes both status quo and actual
problems among São Paulo's *nikkei* society. Although a large number of associations, groups and clubs exist, they experience a kind of recruitment problem connected to a changed relationship to Japan. The contribution of the Fujinkai also
shows that participating meant including oneself and one's organization within
a Japanese community, be it transnational with Japan or regional in São Paulo.
The catastrophes of 3/11 strengthened the relations between actors who worked
together for a common aim: helping Japan by providing donations. Their success lies, therefore, less in the result of this collection and the sum of one million

[4] To facilitate reading, this association for informal social gathering (*shinbokukai*) and its women's
association (*fujinkai*) will be abbreviated to "Fujinkai".

dollars but more in the process of achieving it. Even though actors had organized campaigns for other objectives in Brazil, like the flooding in Rio de Janeiro, the "affected home" Japan added to the equation. Participating and donating meant demonstrating solidarity while identifying as Japanese and being recognized as such by others. The expectation that Japanese were obligated to participate also meant that those who were not present had denied their contribution, meaning they would not identify themselves as Japanese. The *nikkei* community constructed their need to show solidarity not only from a humanitarian perspective but also as an obligation owed to one's own group or family, which further strengthened the process of "ouring" among the self-imposed category of "Japanese".

"Ouring" even extended to contributions from Brazilians without Japanese ancestry. Their favorable image of "Japanese" and therefore motivation to help was developed from the positive stereotypes about the *nikkei* in Brazil and especially in São Paulo. They directly linked the Japanese in Japan to the *nikkei* in Brazil, as a contact recalled the reasoning of one donor in particular:

> There was a woman, who came here right after the events of our campaign started and who we perceived to be a simple domestic maid [...] and she gave twenty Reais [about $ 10]. She said: "No, I insist. This is all I've got, all that I have left at this moment. But I insist. The Japanese are very correct, they are very hard workers, they deserve that we help them" [...]
> "So, you are familiar with Japan?"
> "No, but I know you, and you [plural] are the best."

Helpful as it may have been in the campaign, this form of performative and associated solidarity also posed a problem. The categorization of Japanese through donations became an open field that could and would be contested.

5   Reacting to 3/11: Friction and Conflict

The reaction of the *nikkei* community seems clear and well structured; existing institutions and their representatives assumed responsibility, organized a joint campaign, while individuals and associations contributed with financial donations that were distributed. Help was therefore provided to Japan. But who was allowed to organize campaigns and represent the community? While the variety

of campaigns was remarkable, this also led to problems of differentiating between them. Even those five organizations coordinating the campaign together
used different names: Campaign to Collect for the Victims, Nippon Gambare,
and SOS Japão. Among *nikkei* and non-*nikkei* contributors, there was appar-
ently not only a great desire to help but also a feeling of uncertainty regarding
who was authorized to collect the donations. All official documents and press
releases warned that no one was allowed to gather donations in person, but as
one incident proved, there were those who used ethnicity and this solidarity as a
cover for criminal gains.

Rosário Kazuhaki Yamamoto is a known Japanese-Brazilian con artist who already had posed as a distant relative of Japan's Imperial family to obtain money
during the commemoration of Japanese immigration to Brazil in 2008. In 2011,
he re-appeared in Liberdade wearing a doctor's coat and pretended to be a representative of an association participating in the campaign. He successfully lured
businessmen into sponsoring a fictional fund-raising bazaar (São Paulo Shimbun 2011b). As one victim remarked afterwards, in addition to his doctor's coat,
Yamamoto's look as a Japanese-Brazilian also made him credible, even more so
because he spoke both Japanese and Portuguese. This attempt to capitalize on
the catastrophe was not just seen as an act of a criminal but taken as an "attack on the community" (São Paulo Shimbun 2011b). This reaction shows the
fear that members of the community themselves would besmirch the nationwide
campaign. Yamamoto's example illustrates the warring question of who was allowed to act and by extension speak for and represent the community. Friction
appeared within the community, and in tracing these moments a different perspective of the meaning of 3/11 can be drawn. Two examples of friction among
the community illustrate the variety of meanings.

The biggest contributor to the campaign was the religious sect Seichō-no-ie.
On April 4, its delegation visited the Bunkyo building and delivered a check of
about $ 250,000. The presentation of the donation to the presidents of Bunkyo,
Enkyo, and Kenren followed the routine of other donations: words of gratitude
were spoken and pictures of those involved taken. However, the circumstances
of this contribution were very different. Published pictures show a highly for-

malized setting resembling a press conference with opposing groups and tables –
no other contributors were met like this. Most often, a member of the executive board of Bunkyo received donors, whereas this time, three presidents of
the organizing committee were present. Seichō-no-ie even brought a journalist
from their own journal, thus partly taking control of the presentation afterwards
when they published an account of the meeting on their own websites, adding
their own pictures and a link to an article about their donation (Seichō-no-ie
2011). This case suggests a different reason for participating in the campaign.
It can be concluded that donating in Seichō-no-ie's case meant not only showing one's religious beliefs and solidarity with the Japanese but also presenting
oneself as an influential actor. While Seichō-no-ie is one of the Japanese "New
Religions" (*shinshūkyō*) which have gathered a large following in Brazil, its influence on society is also seen critically (Usarski-Shoji 2008: 3). Bunkyo and other
institutions are normally cautious towards religious organizations, but in this
case, they had to accept the donation and in doing so, also accept Seichō-no-ie
as an important part of campaign and community. Showing solidarity in this
case meant publicity and a marketing possibility because all news media of the
community reported the contribution. In contrast to this, the largest individual
donor who contributed about $ 100,000 chose not to identify himself (Abe-Oi
2011e). Seichō-no-ie's use and distribution of their own material through their
own forms of media also demonstrates the focus on publicity because the published report and pictures concentrate on the gratitude shown by the organizers
as representatives of the *nikkei* community.

In contrast to Seichō-no-ie's example of pursuing one's own agenda through
participating, the example of Liberdade's shop owners association ACAL (Associação Cultural e Assistencial da Liberdade) illustrates the effects of refusal.
If "being Japanese" meant collecting donations, then it became an impetus for
any *nikkei* organization. Those who refused to contribute in any way – not only
financially – to the campaign came under pressure. Shortly after the catastrophes, ACAL's executive board send out a communiqué to its members and associates. It labeled any initiatives to collect donations "hasty", arguing that Japan
had "enough financial and human resources to normalize the situation" (Shiguti

2011b: 4). The message also stated that participating in any campaign would be neither necessary nor wanted by the Japanese government. This drew heavy criticism from the public, the media and ACALs own associates (Shiguti 2011b: 4). Participating was expected, and not surprisingly, some days later, ACAL organized its own one-day campaign in collaboration with the Red Cross (albeit the Brazilian charter).[5]

6 The Meaning of 3/11 for the Japanese Diaspora in Brazil

The Japanese-Brazilian community in São Paulo was a driving force behind efforts to aid Japan after 3/11. It started various initiatives that provided moral and financial support. Nevertheless, to evaluate these contributions is a difficult and contested question. Organizers such as Bunkyo president Kita considered the results a success that "surpassed [their] expectations" (Abe-Oi 2011e), but others criticized both the campaigns and the motives to participate. Rodrigo Meikaru, editor of the journal Mundo OK, described it as "one event here, one event there, but nothing big" and declared that "the Japanese-Brazilian community [...] did not show great efforts to help Japan" (Meikaru 2011: 5). The journalist Nelson Fukai saw a very different picture of "solidarity" when he compared reactions in Japan and Brazil, addressing that participating in Brazil was mainly used for publicity: "Here in Brazil, mainly in the nikkei [sic!] community..., what a shame! Solidarity yes, for those who want to present themselves!" (Fukai 2011). Compared to the sum collected on a global scale, the sum of about one million dollars (about three million Brazilian Reais) from private donations ought to be considered symbolic, as a member of the Bunkyo staff confirmed: "We know that three million [Reais] were not sufficient at all but [meant to] [...] demonstrate a feeling of solidarity" (2011, 021-230).[6] It can be argued that the process of collecting and its impact on the community was more important than the donated sum itself.

---

[5] This short-term solution was criticized because the Brazilian charter of the Red Cross charged ten percent of the collected money as an administration fee. Since late 2011 the Brazilian charter has also been under suspicion for possible embezzlement of funds with the campaign for the victims of 3/11 being among those investigated (Leitão, 2012).

[6] a gente sabe que tres milhões não deu para nada, mas [...] demonstrar um sentimento, ne de soliedaridade (2011, 021-023).

On the one hand, the main campaign cemented the claim of established institutions such as Bunkyō and the *kenjinkai* to represent the *nikkei* community and made them more visible. Participating meant constructing one's identity as being Japanese. The public campaigning for donations and its "ouring" as an expression of ethnic solidarity also meant mobilizing broader audiences, both non-*nikkei* and *nikkei*, which resulted in the "strengthening [not only] of local relations but also of relations inside the community as well as with local [Brazilian] society", as a Bunkyō contact stated.[7] Participating in the campaign also had an impact on power relationships, either gaining (positive) publicity or sanctions from a refusal. While the act of donating was publicly acknowledged (and partly self-presented), the amount of money could also serve as a confirmation of one's own relevance and power in contrast to others. It may be concluded that the catastrophes of 3/11 were a tragic event in which the Brazilian *nikkei* diaspora experienced and constructed its connection to Japan: on the one hand, internally through its own efforts with campaigns, and on the other hand, externally through the attention of non-*nikkei* society, which supported and expected this symbolic show of solidarity with Japan.

This leads to a general question of how diasporic communities respond to disasters in their (ethnic) homeland. As shown in this article, disasters can be seen as experiences of a connection to one's homeland, tragically conforming to Safran's idea of relating and restoring. Regardless of geographic locations, modern media supplies pictures and other impressions that trigger reactions. Solidarity and its various forms of expressions, varying from financial donations to public demonstrations, become an externally and internally expected reaction. Identity is constructed from the periphery through internal "ouring" but may be confirmed by external opinions.

The catastrophes of 3/11 had global repercussions on the Japanese diaspora. It may be argued that they became part of the diaspora's history and in this sense ought to be remembered. In March 2012, two *nikkei* organizations supported by Mitsubishi and the city of São Paulo began to plan a memorial. Twenty thousand

---

[7] "um fortalecimento das relações locais mas também dessas relações dentro da comunidade como também da sociedade local" (2011, 002).

tree seedlings, representing the number of assumed victims, are to be planted in a park in the Eastern district of São Paulo (Jornal Nippak 2012). So far, donations are still being collected, but naming this memorial the "Kizuna Park Brazil Japan" (Bosque Kizuná Brasil Japão) shows a close orientation toward the discourse of Japanese politics and of "bonds between people" (*kizuna*) in the reactions to the catastrophes in Japan (Tagsold, 2012). It remains to be seen how the catastrophes of 3/11 will be seen and remembered not only in São Paulo but also in other diasporic *nikkei* communities.

Literature

ABE OI, Célia. (2011a): "Manifesto de Soliedaridade ao Japão da comunidade nipo-brasileira" [Manifest of solidarity with Japan from the Japanese-Brazilian community]; http://www.bunkyo.bunkyonet.org.br/index.php''option=comcontent&view=article&id=1033%3Amanifesto-de-solidariedade-do-japao-da-comunidade-nipo-brasileira&catid=84%3Aterremoto-2011&Itemid=122&lang=br. (accessed October 12 2014)

ABE OI, Célia. (2011b): "Do Brasil, a Mensagem de Solidariedade aos Japoneses" [A Message of Solidarity to the Japanese from Brazil]; http://www.bunkyo.bunkyonet.org.br/index.php''option=comcontent&view=article&id=1035%3Ado-brasil-a-mensagem-de-solidariedade-aos-japoneses&catid=84%3Aterremoto-2011&Itemid=122&lang=br. (accessed October 14, 2014)

ABE OI, Célia. (2011c): "Cruz Vermelha do Japão receberá a arrecadação da campanha em prol das vítimas do terremoto" [Red Cross of Japan Will Receive the Collection of the Campaign on Behalf of the Victims of the Earthquake]; http://www.bunkyo.bunkyonet.org.br/index.php''option=comcontent&view=article&id=1047%3Acruz-vermelha-do-japao-recebera-a-arrecadacao-da-campanha-em-prol-das-vitimas&catid=84%3Aterremoto-2011&Itemid=122&lang=br. (accessed October 14, 2014)

ABE OI, Célia. (2011d): "Vila Santa Izabel na Campanha de Arrecadação às Vítimas do Terremoto no Japão" [Vila Santa Izabel in the Campaign to Collect for the Victims in Japan]; http://www.bunkyo.bunkyonet.org.br/index.php''option=comcontent&view=article&id=1101%3Avila-santa-izabel-na-campanha-de-arrecadacao-as-vitimas-do-terremoto-no-japao&catid=84%3Aterremoto-2011&Itemid=122&lang=br. (accessed October 15, 2014)

ABE-OI, Célia. (2011e): "Resultado final da Campanha em prol das vítimas do terremoto no Japão" [Final Result of the Campaign for the Victims of the Earthquale in Japan]; http://www.kenren.org.br/noticias/not%C3%ADcias-relacionadas/item/67-result

ado-final-da-campanha-em-prol-das-v%C3%ADtimas-do-terremoto-no-japão. (accessed October 29, 2014)

ADACHI, Nobuko. 2004. "Japonês: A Marker of Social Class or a Key Term in the Discourse of Race". In: *Latin American Perspectives*, 31 (3): pp. 48–76.

ADACHI, Nobuko. 2006. "Theorizing Japanese Diaspora" In: Adachi, Nobuko. (ed.): *Japanese Diasporas: Unsung Pasts, Conflicting Presents and Uncertain Futures*. New York: Routledge, pp. 1–22.

BELTRÃO, Kaizō et al. 2008. "Haíne: Raízes [Roots]." Curitiba: Associação Brasileira de Dekasseguis.

BLOG DO PLANALTO. 2011. "Presidenta Dilma coloca o Brasil à disposição do governo japonês após terremoto" [President Dilma Places Brazil at Disposal of the Japanese Government after Earthquake]; http://blog.planalto.gov.br/presidenta-dilma-coloca-o-brasil-a-disposicao-do-governo-japones-apos-terremoto. (accessed July 26, 2014)

BRODY, Betsy Teresa. 2002. *Immigration, Ethnicity, and Globalization in Japan*. London: Routledge.

FUKAI, Nelson. 2011. "O marketing de disgraça" [Shameful Marketing]. In: *Mundo OK*: 4.

HANDA, Tomoo. 1987. *O imigrante Japonês – História de sua vida no Brasil* [The Japanese Immigrant - History of His Life in Brazil]. São Paulo: Centro de Estudos Nipo-Brasileiros.

HARADA, Kiyoshi. 2010. "A Presença do Nikkei no Cenário Nacional" [Presence of Nikkeis on the National Scenario]. In: *Various. Centenário: Contribuição da Imigração Japonesa para o Brasil Moderno e Multicultural* [Centenário: Contributions of the Japanese Immigration to Brasil Modern and Multicultural]. São Paulo: Paulo's Comunicação e Artes Gráficas, pp. 223–235.

HOUAISS, Antônio and; VILLAR, Mauro de Salles. 2001. *Dicionário Houaiss da Língua Portuguesa*. Rio de Janeiro: Objetiva.

JORNAL NIPPAK. 2012. "Convite ABIJA e OISCA – Plantio de árvores dia 17/03/2012" [Invitation ABIJA and OISCA – Planting of Trees at 17/03/2012]; http://www.portalnikkei.com.br/convite-abija-e-oisca-plantio-de-arvores-dia-17032012/. (accessed August 19, 2014)

LEITÃO, Leslie. 2012. "S.O.S. Cruz Vermelha" [S.O.S. Red Cross], VEJA 2281, 8.8.2012, pp. 120–123.

LESSER, Jeffrey. 2007. *A Discontented Diaspora: Japanese Brazilians and the Meanings of Ethnic Militancy, 1960–1980*. Durham: Duke University Press.

MEIKARU, Rodrigo. 2011. "Senso de solidariedade" [A Sense of Solidarity]. In: *Mundo OK*: 4.

OKADA, Norio et al. 2011. "The 2011 Eastern Japan Great Earthquake Disaster: Overview and Comments". In: *International Journal of Disaster Risk Science*, 2 (1): pp. 34–42.

REIS, Maria Edileuza Fontanele. 2002. *Brasileiros no Japao: o elo humano das relacoes bilaterais* [Brazilians in Japan: The Human Tie of the Bilateral Relationship]. São Paulo: Kaleidus-Primus. 2nd revised edition.

ROSSO, Paulo Sergio. 2007. "Solidariedade e direitos fundamentais na Constituição Brasileira de 1988" [Solidarity and Fundamental Rights in the Brazilian Constitution of 1988]. In: *Revista Eletrōnica do Centro de Estudos Jurídicos/CEJUR*, 1 (2): pp. 201–222.

SÃO PAULO SHIMBUN. 1995a. "Morte e sofrimento no rastro do terremoto" [Death and Suffering in the Wake of the Earthquake]. 18.1.1995.

SÃO PAULO SHIMBUN. 1995b. "A soliedaridade da comunidade" [The solidarity of the Community]. 20.1.1995.

SÃO PAULO SHIMBUN. 1995c. "Brasileiros estão sendo atendidos em Osaka" [Brazilians Are Treated in Osaka]. 2.2.1995.

SÃO PAULO SHIMBUN. 1995d. 04.02.1995.

SÃO PAULO SHIMBUN. 2011a. "Brasil doa US$ 500 mil e Japão agradece" [Brazil Donates 500.000 US$ and Japan Gives Thanks]. 22 March; http://www.saopauloshimbun.com/sitebr.php/conteudo/show/id/1079/menu/31/cat/112. (accessed June 2, 2014)

SÃO PAULO SHIMBUN. 2011b. "Estelionatário volta a atacar a comunidade usando o nome do São Paulo Shimbun" [Trickster Returns to Attack the Community Using the Name of SãO Paulo Shimbun]. August 29; http://www.saopauloshimbun.com/sitebr.php/conteudo/show/id/1450/menu/33/cat/115. (accessed October 15, 2014)

SAFRAN, William. 1991. "Diasporas in Modern Societies: Myths of Homeland and Return". In: *Diaspora*, 1 (1): pp. 83–99.

SEICHŌ-NO-IE. 2011. "SEICHO-NO-IE DO BRASIL faz entrega de doação que será encaminhada às vítimas do terremoto e tsunami do Japão" [Seichō-No-Ie Delivers the Donation Which Will Be Sent to Victims of the Earthquake and Tsunami in Japan]; http://www.sni.org.br/SEICHO-NO-IE-DO-BRASIL-faz-entrega-de-doacao-que-sera-enca minhada-as-vitimas-do-terremoto-e-tsunami-do-Japao.asp. (accessed August 29, 2014)

SHIGUTI, Aldo. 2011a. "O pânico maior é por parte dos estrangeiros, diz Hashimoto" [The Bigger Panic Comes from the Foreigners, Says Hashimoto]. In: *Jornal Nippak*, 17-23.03.2011.

SHIGUTI, Aldo. 2011b. "Acal se retrata por informe 'equivocado' sobre campanhas" [Acal Recants "Mistaken" Information about Campaigns]. In: *Jornal Nippak*, 31.3.-06.04.2011.

SPIVAK, Gayatari. 1985. "The Rani of Simur". In: Barker, Francis et al (eds.): *Europe and its Others*. Colchester: University of Sussex.

TAGSOLD, Christian. 2012. "Kizuna: Das Schriftzeichen des Jahres 2011 als Antwort auf das gefühlte Auseinanderbrechen der Gesellschaft" [Kizuna: The Kanji of the Year 2011 as an Answer to the Angst That Japanese Society Will Soon Break Apart]. In: *Japan Jahrbuch*: pp. 309–328.

TAKEZAWA, Yasuko. 2002. "Nikkeijin and Multicultural Coexistence in Japan." In: Hirabayashi, Lane Ryo: *New Worlds, New Lives: Globalization and People of Japanese Descent in the Americas and from Latin America in Japan*, Stanford: Stanford University Press, pp. 310–330.

USARSKI, Frank; and SHOJI, Raphael. 2008. "Japanese Religions in Brazil". In: *Japanese Journal of Religious Studies*, 35 (1): pp. 1–12.

About the Authors

NIKO BESNIER is Professor of Cultural Anthropology at the Department of Sociology & Anthropology, University of Amsterdam

PETER BERNARDI is a PHD candidate at the Department for Modern Japan, Heinrich Heine University Düsseldorf

RUTH MARTIN is Research Associate at the Europe Japan Research Centre, Oxford Brookes University

ANDREAS NIEHAUS is Associate Professor at the Department for Japanese Language and Culture, Ghent University

CHRISTIAN TAGSOLD is Associate Professor at the Department for Modern Japan, Heinrich Heine University Düsseldorf

JUTTA TEUWSEN is a Research Fellow at the Department for Modern Japan, Heinrich Heine University Düsseldorf

TINE WALRAVENS is Teaching and Research Assistant at the Department of Languages and Cultures Department / Japanese Language and Culture, Ghent University

www.ingramcontent.com/pod-product-compliance
Lightning Source LLC
Chambersburg PA
CBHW052013270326
41929CB00015B/2902